THE ENTANGLED MIND

The Green Knight,
Black Holes,
and Quaternity

Wes Jamroz

Troubadour Publications

The Entangled Mind

The Green Knight, Black Holes, and Quaternity

Editing: Dominique Hugon, Patrick Barnard

Cover illustration: Sandra Viscuso
                              (viscuso.sandra@gmail.com)

Montreal, QC, Canada

TroubadourPubs@aol.com
http://www.troubadourpublications

ISBN: 978-1-928060-17-8

Every particle of dust is a cup wherein

all the world can be seen.

(Gharib-Nawaz)

Table of Contents

A New Science

Physics performs experiments involving an elaborate teamwork and a highly sophisticated technology, whereas mystics obtain their knowledge purely through introspection, without any machinery, in the privacy of meditation.

(Fritjof Capra)

Today's science is in crisis. Even though we are being bombarded daily by headlines announcing "breakthroughs" in cosmology, quantum mechanics, and elementary particle physics, the most materialistic branch of science, i.e., physics, is at a dead-end.

It is believed that physics provides a model for understanding the universe we live in. The most attractive feature of physics is its power of prediction. If we have enough information about a system, we can know how that system will evolve by applying laws and theories of physics. The reverse is also true. If we know the state of the system now, we can run the process backward to figure out how the system got to its present state. These two concepts are known as determinism, i.e., we can predict the future and we can read the past. This is pretty much the foundational core of physics. However, these basic features of physics have met a serious challenge with the development of quantum mechanics, i.e., that part of physics that works with the most elementary particles of matter. The laws discovered by quantum mechanics seem to indicate that even the simplest forms of matter exhibit some forms of ...

consciousness. And this is a big problem because the term "consciousness" does not belong to the vocabulary of hard-core physics. The overall situation is even more complicated because the term "consciousness" is not satisfactorily defined by any soft science. Therefore, we are faced with quite a peculiar puzzle. On the one hand, hard sciences would have to come to terms with "consciousness." On the other hand, soft sciences (philosophy, psychology, sociology, etc.) would have to provide a more satisfactory definition of what "consciousness" really is. As we can imagine, this situation has created quite a challenge for the entire scientific community.

The general literature regarding this challenge has featured two seemingly contradictory approaches to understanding the universe: a scientific method and a perceptive one (often referred to as "mystical"). In the past, there were many attempts to bridge these two approaches. The most successful among them were those that tried to explain some scientific discoveries by referring them to various ancient mystical texts. For example, the celebrated *Tao of Physics* by Fritjof Capra claimed many similarities between some of the fundamental notions of major Eastern religions and modern physics. This claim has encouraged many other writers to pursue such an approach in which new scientific discoveries are discussed in the context of mystical texts. However, the overall pattern of such discussions has been the same. Namely, the latest scientific discoveries were traced back to ancient texts. In this way, scientific discoveries were used to explain some seemingly obscure pronouncements of ancient writings. Yet, it was science that was leading and stimulating these discussions.

The aim of *The Entangled Mind* is to show that the perceptive approach should not be limited to commentary on the latest discoveries. The perceptive approach can effectively provide solutions for the current impasse in science. In other words, this book marks a new role for the perceptive approach. This book attempts to demonstrate that the perceptive approach may serve as a guide for the further advancement of science. At the same time, this book illustrates how the human mind has been gradually prepared for such a transition through poetry, songs, fairy tales, and certain games.

The Entangled Mind describes how the so-called "mysticism" gradually becomes a new science and how deterministic science gradually turns into a dogmatic religion.

The Elephant in the Dark Room

You can discover truth only if you are willing to give
your whole mind and heart to it, not a few moments of
your easily spared time.

(Jiddu Krishnamurti)

"Consciousness" remains a puzzling and controversial term despite millennia of philosophical and scientific analyses, definitions, explanations, and debates. In its most common understanding, consciousness is defined as awareness of internal and external existence. Sometimes, it is thought of as synonymous with the mind or an aspect of the mind. For others, this term refers to individual awareness of unique thoughts, memories, feelings, sensations, and environments. Often it is stipulated that there are different types of consciousness, e.g., that of humans, animals, or even the whole universe. However, the only widely agreed notion about the topic is that ... consciousness exists.

Scientists and philosophers have proposed countless hypotheses of what consciousness is and how it arises. For example, *panpsychism* contends that all creatures and even inanimate matter possess consciousness. Conversely, hardcore *materialists* insist that not even humans are all that conscious. Then *solipsism* holds that knowledge of anything outside one's mind is unsure; the external world and other minds cannot be known and might not exist outside the mind. So far, scientists and ethicists who study the issue say that nobody has created consciousness in a laboratory. And, according to an idea called *integrated information theory*, consciousness is a product of how densely neuronal networks are connected across the brain. The more neurons interact with one another, the higher the degree of consciousness – a quantity known as *phi*. If *phi* is greater than zero, the organism is considered conscious.

Interest-specific biases towards consciousness further complicate the overall situation. As a result, there are different definitions proposed by psychologists, linguists, anthropologists, philosophers, biologists, and sociologists. This kind of approach has not helped in any meaningful way advancement in the understanding of consciousness. There is no common denominator that would link all these various descriptions into one coherent definition of what consciousness is. An overall model that would encompass all those diverse and fragmented types of understanding into a single and comprehensive framework – is missing.

The following statements by some of the researchers in consciousness illustrate the lack of such a framework:[1]

- Ludwig Wittgenstein, one of the most influential philosophers of the 20th century, said that if a lion could talk, we wouldn't understand it. In this way, he indicated our inability to communicate with other types of consciousness.
- Christof Koch, a neuroscientist, says that as long as we lack a sort of consciousness meter, any theories of consciousness will remain in the realm of pure speculation. Consequently, he hopes that one day we all get brain implants with Wi-Fi so we can combine our minds through a kind of high-tech telepathy.
- Philosopher Colin McGinn, on the other hand, suggests

[1] "How do I Know I'm Not the Only Conscious Being in the Universe?" John Horgan, *Scientific American*, Special Collector's Edition, Winter 2022, p. 32.

resolving the problem through a technique that involves "brain splicing." In this way, it might be possible to "measure" our consciousness by transferring bits of a person's brain into another person. In this "technique," the other person would serve as a meter of our consciousness.

- Steven Laureys, a neurologist, thinks it is futile to try to identify consciousness in any lab-maintained brain. He believes that such laboratory experiments are useless as long as we do not understand what consciousness is.
- Another philosopher, Philip Goff, posits an entirely different belief. He suggests that consciousness may come from an alien programmer, or perhaps it pervades our universe, not just our brains but all things.

To complete the overall picture, let's look at what Roger Penrose has to say about the subject. Penrose is one of the most influential physicists of our time. In 2020, he received the Nobel Prize in Physics for his work on black holes. Here are some of his thoughts on consciousness:[2]

- I do not think that it is wise, at this stage of understanding, to attempt to propose a precise definition of consciousness, but we can rely, to good measure, on our subjective impressions and intuitive common sense as to what the term means and when this property of consciousness is likely to be present.
- I more or less know when I am conscious myself. /.../
- To be conscious, I seem to have to be conscious of something, perhaps a sensation such as pain or warmth or colorful scene or musical sound; or perhaps I am conscious of a feeling such as puzzlement, despair, or happiness; or I may be conscious of the memory of some

---

[2] *The Emperor's New Mind*, Roger Penrose, Oxford University Press, New York, 1989, p. 406.

past experience, or of coming to an understanding of what someone else is saying, or a new idea of my own, or I may be consciously intending to speak or to take some other action such as get up from my seat.
- I may be asleep and still be conscious to some degree, provided that I am experiencing some dream; or perhaps, as I am beginning to awake, I am consciously influencing some dream; or perhaps, as I am beginning to wake, I am consciously influencing the direction of that dream.
- I am prepared to believe that consciousness is a matter of degree and not simply something that is either there or not there.
- I take the word "consciousness" to be essentially synonymous with "awareness," ... whereas "mind" and "soul" have further connotations, which are a good deal less clearly definable at present. /.../
- We shall be having enough trouble with coming to terms with "consciousness" as it stands, so I hope that the reader will forgive me if I leave the further problems of "mind" and "soul" essentially alone! /.../
- I should be doubtful that a worm or insect - and certainty not a rock - has much, if anything, of this quality, but mammals, in a general way, do give me an impression of some genuine awareness. From this lack of consensus, we must infer, at least, that there is no generally accepted criterion for the manifestation of consciousness.

Although expressed in plain language, Penrose's meditation adequately illustrates the overall state of understanding of consciousness among the leading representatives of today's academia. Furthermore, it is apparent that there is a lack of a framework within which "consciousness" could be defined. All of these indicate that the approaches applied so far are based on fragmented views, which are insufficient for a satisfactory definition of consciousness. A major conceptual adjustment is needed to address this challenge in an effective manner.

As long as "consciousness" and its overall modus operandi is not understood, scientists will be locked up in a dark room with that proverbial elephant of unknown shape.

The State of Today's Physics

If you thought that science was certain - well,
that is just an error on your part.

(Richard P. Feynman)

Today's physics faces challenges on several fronts. They are related to the nature of the big bang, elementary particles, dark matter, black holes, and some seemingly mysterious aspects of quantum mechanics. Interestingly, all these challenges boil down to the same problem, which this book attempts to identify and solve. First, however, let's briefly summarize these challenges.

<u>The Big Bang</u>

Until the 20th century, scientists assumed that the universe was static and remained unchanged throughout eternity. Then, in 1915, Einstein developed the general theory of relativity. This theory describes how gravity acts across the fabric of spacetime. Einstein was puzzled to find that his theory indicated that the cosmos must either expand or contract. To preserve a static universe, he had to make some changes in his model by adding a factor called the "cosmological constant." This was strictly a "cosmetic" adjustment of the theory as there was no evidence to justify it; it was introduced simply to comply with the *belief* in a static universe.

A few years later, astronomers Georges Lemaître and Edwin Hubble made the startling discovery that galaxies were zipping away from each other. The observations confirmed that the universe was not static at all; the universe was expanding. This means that sometimes in a very distant past, there must have been a time when everything in the universe was close together. This assumption led to the concept of the big bang.

According to today's science, the physical universe was created as the result of the big bang. The idea of the big bang is based on a theory which predicts that all matter can be compressed into a region of infinitely small volume, a sort of one-dimensional dot. Any mathematical theory breaks down at this point because no object can be defined at such a condition. This condition is what physicists call a singularity. Consequently, physicists assume that the universe was born from that one-dimensional dot. In other words, the universe must have begun with a singularity. This means that no mathematical model can be applied to any object or event before the big bang. As far as science is concerned, before the big bang – there was … nothing.

Such an understanding of the universe is hard to swallow for those who intuitively sense that something is missing from that picture. Unfortunately, there is insufficient terminology and conceptual understanding that would provide a common platform allowing physicists and non-physicists to engage in a meaningful and constructive discussion. Poets and artists, therefore, engaged in that discussion on their own terms. For example, resentment of the strictly rational and contra-intuitive concept of the big bang is nicely reflected in a poem by Marie Howe, a contemporary American poet.[3] The poem, entitled "Singularity," was inspired by Stephen Hawking's description of the big bang.

---

[3] https://www.youtube.com/watch?v=on7UECZq_nA (November 18, 2020).

Do you sometimes want to wake up to the singularity
we once were?

so compact nobody
needed a bed, or food or money –

nobody hiding in the school bathroom
or home alone

pulling open the drawer
where the pills are kept.

For *every atom belonging to me as good*
Belongs to you. Remember?
There was no *Nature.* No
them. No tests
to determine if the elephant
grieves her calf   or if

the coral reef feels pain. Trashed
oceans don't speak English or Farsi or French;

would that we could wake up   to what we were
– when we *were* ocean   and before that
when earth was sky, and animal was energy, and rock was
liquid and stars were space and space was not

at all – nothing

before we came to believe humans were so important
before this awful loneliness.

Can molecules remember it?
what once was?   before anything happened?

can our molecules remember?

No I, no We, no one. No was
No verb   no noun yet
but only a tiny tiny tiny tiny dot brimming with

is is is is is

All   everything   home

"Can molecules remember it?" – the poet is asking an adequate question. In the following chapters, we will attempt to find the answer.

Quantum effects

To classical physicists, the world was ultimately rational. Things had to make sense. They had to be quantifiable and expressible through a logical chain of cause-and-effect interactions, from what we experience in our everyday lives all the way to the depths of reality. However, according to quantum mechanics, we have no right to expect any such order or rationality. At its deepest level, nature does nor need to follow any of our expectations of well-behaved determinism. Quantum mechanics divided the world into two realms, the familiar classical world, and the unfamiliar quantum world.

The fundamental idea of quantum mechanics is the concept of state. Every particle is basically in the form of a wave. Consequently, the state of a particle is completely characterized by the so-called wave equation. In accordance with quantum mechanics, one can only calculate the relative *probability* that a particle is at a particular location at a given time. A definite value of the location may be obtained only by measurement. When a quantum measurement is performed, the probabilistic state of the particle is transformed into a fact. This transformation is called the "collapse" of the wave function. However, the "collapse" disturbs the particle and thus alters its state. After its position is experimentally determined, the particle will no longer be in the state given by the wave equation. This means that there will always be a discrepancy between theoretical calculations and the measured state of particles. This discrepancy is known as the "quantum measurement problem." This is an unsolved problem because it is impossible to observe directly how the wave function collapses; the process of the collapse itself remains a mystery. It was concluded, therefore, that the result of an experiment is no longer absolute; it is somehow dependent on the experimenter. Because it is the experimenter who "creates" the outcome of the experiment. Following their deterministic *belief*, physicists further assumed that the measurement affects not only quantum particles but must also apply to all physical objects. This assumption was the source of a series of ridiculous paradoxes, such as the famous Schrödinger's cat, supposedly both dead and alive, until someone looks in the box. Or Einstein's box with gunpowder which is simultaneously "not-yet-exploded" and "already exploded."

Secondly, quantum mechanics indicates that the more accurately a particle's state is measured, the more the measuring process itself disturbs it. This so-called "uncertainty principle" forbids any measuring process from extracting all the information about a particle. This principle is a fundamental law of quantum mechanics. Therefore, in addition to probability, quantum mechanics also introduces an element of *uncertainty*. If these properties apply to all forms of matter, then one cannot precisely measure the present state of the universe. Neither is it possible to predict precisely future events. The discovery of quantum mechanics signaled the end of a completely deterministic universe.

Yet, there are further complications. Namely, it was discovered that the laws governing the quantum world indicate that some form of awareness might be involved in the behavior of such elusive objects as photons and electrons. For example, the effect known as quantum entanglement demonstrates that two particles can be made aware of each other. The entangled particles seem to know about the states of their entangled partners. Quantum entanglement essentially allows two particles to behave as one, regardless of how far apart they are. As a result, the measurements performed on one particle seem to be instantaneously influencing the other particle that is entangled with it. Einstein referred to this as a "spooky action at a distance." The spookiness of this effect lies in the fact that there is no physical link between the entangled particles.

How is this possible? How can particles be linked without a link? We will find out.

Elementary particles

According to quantum mechanics, the fundamental building blocks of nature are some spooky waves of various shapes that spread throughout all of space. These substances are called fields. A physical field can be thought of as having energy at each point in space and time. In certain circumstances, the waves are manifested as particles. For example, the waves of the electromagnetic field give rise to particles called photons. So, photons are derivatives of the electromagnetic field.

The same process is at play for all other elementary particles that we know of. Elementary particles such as photons and electrons have known values for their properties, such as energy, momentum, and spin. They are often considered to be single point particles of zero size. However, their mass is not defined. If a particle had zero size and some mass, then this would imply an infinite density. This means that elementary particles are not solid objects in the physical sense. This is why it is a meaningless question to ask what the size of a photon is. This, in turn, implies that the particles are forms of various fields. Every particle is a tiny ripple of an underlying quantum field.

So, according to modern physics, space is filled with a collection of fields. But these are quantum fields. Therefore, their states are inherently probabilistic and uncertain. Consequently, it is impossible to have, for example, a point of precisely zero energy within a quantum field. This would violate the principle of uncertainty. So, even if a quantum field remains in its minimum energy state, it will have some energy. Such a state with minimum energy corresponds to what is known as a quantum vacuum. Sometimes, the quantum vacuum is referred to as "nothingness." Yet, a quantum vacuum does not mean that there is nothing. Instead, a quantum vacuum contains fleeting fields and particles that pop into and out of it. According to quantum mechanics, such constant fluctuations in energy can spontaneously create mass not just out of thin air, but out of … nothing. Professor David Tong of the University of Cambridge admits: "This is difficult. Many decades after quantum field theory was constructed, we are still a long way from understanding all the subtleties that it contains."[4]

So, is it possible to create something from nothing?

All this wordplay with "vacuum" and "nothingness" is to avoid stating the obvious: physicists do not understand how matter is created. As we will see in the following chapters, something else is needed to resolve this problem.

Dark matter

Today's understanding of the cosmos is also in a bit of turmoil. Cosmologists understand how stars are formed, how they are burnt, and how they die. But cosmologists have a problem. Despite recent advances in astrophysics and astronomy, they still don't understand exactly how galaxies can exist.

The overall process is seemingly simple. Galaxies are collections of stars held together by gravity. Like our solar system, stars follow their predetermined trajectories. Depending on their traveling speed, stars require different gravity forces to keep them in their orbits. This means that the stars that are moving faster require a stronger force to hold them in their orbits. However, the measurements of orbital speeds indicate that the stars are moving so fast that … galaxies should be torn apart. It seems that there is not enough gravity to explain the galactic dynamics.

The accepted explanation for this observational conundrum is the existence of another unknown substance that would make up for the missing gravity. This hypothetical substance is called "dark matter." It is called "dark" because it does not emit light or any known form of radiation. Like "nothingness," this substance can't be seen by any of the instrumentation that astronomers use to observe the cosmos. Yet, dark matter is needed to stabilize the clusters of galaxies. However, a more detailed analysis of that concept concluded that it is not enough to have dark matter. In the late 1990s, the Hubble telescope discovered that the universe's expansion was accelerating. To account for that growing acceleration, it was proposed that some unknown form of "dark energy" was responsible for pushing galaxies apart more strongly than gravity pulled things together. Dark energy was added to account for the rate at which the universe expands. The amount of those dark substances needed to explain the dynamics of the cosmos is quite surprising or even shocking. According to estimates, dark matter comprises 26.8 % of the universe; dark energy 68.3 %.

All visible matter (marked as "atoms" on the following diagram) which includes planets, stars, galaxies, etc., constitutes only 4.9 % of what is needed to explain the dynamics of the cosmos. This means that 95% of the stuff from which the universe is made – is still unknown!

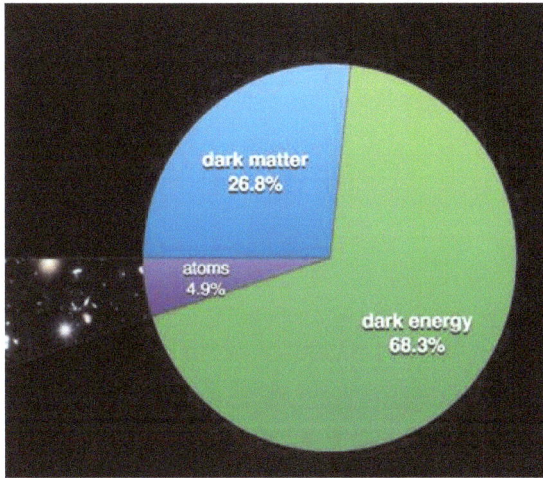

Dark matter and dark energy[5]

Although dark matter is a central part of the standard cosmological model, it's far from being understood. There continues to be nagging mysteries about this stuff – as there is not the slightest experimental evidence of it. As far as dark matter is concerned, physicists are faced with another puzzle that is still waiting for a solution.

So, what is this mysterious "dark matter"? Again, something else is needed to come to terms with dark matter.

Black holes

There is hardly anyone who does not know of the concept of black holes. They have always been a fascinating topic to discuss and wonder about.

To cosmologists, black holes are monsters with a gravity so strong that they can consume stars, wreck galaxies, and

[5] https://physics.stackexchange.com/questions/479098/how-do-they-know-the-numbers-of-the-energy-pie-chart-of-the-universe

imprison even light. At a black hole's center, matter shrinks to infinite density and the known laws of physics break down. At the edge of a black hole – time seems to stop.

The concept of black holes is not as new as one would think. It was already considered in Newton's time. It was then that the possibility of gravity capturing light was first discussed. However, the popularity of this concept grew among scientists when it was discovered that black holes were a consequence of Einstein's general theory of relativity. According to Einstein's theory, the gravitational pull of a compressed mass could become so strong that nothing would escape from its vicinity. In 1916, the idea was first fully conceptualized by Karl Schwarzschild. Schwarzschild found that when a mass remains fixed and its radius decreases gradually, the density of the matter comes to a certain point where even light cannot escape from it. The radius at this point is called the "Schwarzschild radius," and the corresponding surface of the compressed matter is known as the "event horizon." The "event horizon" is a boundary of no return: if you cross it, you won't come back. This idea was a source of inspiration for many writers of science fiction.

At first, scientists did not believe such conditions could occur naturally in the universe. It was only in the 1960s that the existence of black holes was proven theoretically by Stephen Hawking and Roger Penrose. At that time, it was determined that when a star is large enough – it collapses into a black hole. Afterwards, the overall concept became even more complicated. In 1974 Hawking showed that black holes were neither truly black nor eternal. Over time, a black hole would leak energy and elementary particles. As a result, a black hole would shrink, become increasingly hot, and finally explode. In the process, all the mass that had fallen into the black hole would be returned to the outer universe as a sputtering of particles and radiation.

In other words, the process is a manifestation of the cosmic resurrection of matter.

Today, scientists speculate about the properties of resurrected matter. They ask themselves this question: is there any link between the particles that ended up in a black hole and those re-injected back into spacetime? This question led to a further development of the entire concept. Recently, it was suggested that a black hole does not spit out particles randomly. The radiation would start out as random, but as time went on, the emitted particles would become more and more correlated with those that had come out earlier. This suggestion assumes that those newly emitted particles were already entangled with their mates in the black hole.

This leads to a much more fundamental question, which is important to our discussion. Namely, what would be the reason for the resurrection of the matter if the "swallowed" and "spitted out" particles were in the same relationship? Or, if phrased in a more illustrative form: why would a cat spitted out of a black hole be the same as that which fell into it?

We will also find the answer to this question.

***

Quantum mechanics has brought scientists to a door leading to a new branch of scientific inquiry that we will call perceptive science, i.e., an inquiry that reaches beyond the deterministic approach. However, opening this door and entering it will require scientists to make a "conceptual quantum jump" over the limitations imposed by classical determinism. Scientists' *belief* in the utterly deterministic nature of the world served them very well for hundreds of years. However, today's physicists will have to abandon their classical and simplistic determinism to overcome the obstacle they themselves have discovered. They will have to accept that "consciousness" is the fundamental parameter without which it is impossible to overcome the present impasse that physics has been locked in for the last hundred years.

It is probably not so surprising that these two subjects, consciousness and science, are two aspects of the same but much larger framework. Within that larger framework, comprehension of consciousness may help to understand quantum mechanics. At the same time, quantum mechanics may help clarify the entire field of consciousness.

I Want to Earn Her Favor

Words belong to the language of science.
Symbols, to the verbs of the mystical.

(Abu Bakr al-Shibli)

Interestingly, the overall concept of quantum mechanics and consciousness and the way out of the current impasse have been gradually disclosed within seemingly unrelated series of songs, poems, and tales. Although the main objective of those literary devices was to guide humanity along its evolutionary path, their secondary and tertiary impacts have influenced the development of science.

This sort of "impact" upon society is one of the most difficult concepts to grasp. It may take several centuries before their effect on the development of societies is recognized. As always, in its first appearance, such an impact is customized and limited to a specific region and time. At this point, it may be interesting to go back several hundred years in the history of western society and look for hints about impacts which influenced the development of modern physics.

As far as western society is concerned, it all started in 11th century Provence. Namely, the first traces of the overall concept of the existence of quantum-like states were brought to light by the Troubadours. Yes, that's correct. It may be said that the concept of the quantum field was first indicated in the love songs of the Troubadours.

The Troubadours seemed to appear from nowhere, and when they appeared in 11ᵗʰ century Provence, their form was fully developed. In its overt form, the songs of the Troubadours seemed to venerate woman. She was the poet's ideal mistress whom he worshipped from afar without any hope of obtaining her favor. Between the lines of these love poems, however, there was a thinly disguised indication of the existence of another realm. The songs implied that there was a possibility to access that other realm. The *lady* was the means leading to the entanglement of the human mind with that realm.

It is interesting to note another aspect of those spiritually charged love poems. Namely, the appearance of the Troubadours' poems marked the beginning of a new civilization – Western Europe. Similarly, the Homeric poems marked the beginning of historical Greece. So, it looks like these sorts of impacts affect not only a person or a group of people. They carry a significant charge that influences the entirety of human history.

***

Historically, Guillaume de Poitiers (1071 - 1127), the Duke of Aquitaine and Gascony, was the first western Troubadour. Here is one of Guillaume's songs that may serve as a representative sample of Troubadours' love poetry. It is entitled "Very happily, I begin to love" (*Molt jauzions mi prenc en amar*).[6] The version presented here is a slightly edited translation of that song:

---

[6] https://lyricstranslate.com/en/molt-jauzions-mi-prenc-en-amar-very-happily-i-begin-love.html

Very happily I have fallen in love,
Love which will bring much Joy.
Since I desire happiness
I must aim for the best,
and I know that I love the Purest,
the Highest in the entire world.

I should not brag about it,
neither do I dare to praise myself;
but if ever could my joy blossom
it should be that one
that rises above all others
just as the Sun reigns over the skies.

No one could ever describe it,
for neither want, wish,
thought nor imagination
could encompass such a Joy,
and a thousand years is not enough
to express its praise adequately.

Every other joy must lower itself
and all royalty obey my Lady
because of Her wisdom
and of Her beauty;
for whoever can win Her love
will gain immortality.

Her love can heal the sick,
and She changes the greedy into the generous,
an ignoramus into a wise,
an ugly into one handsome,
a ruffian into a courtier,
and the gloomy into the bright faced.

Since nobody can find such a Lady,
so no eye can see, nor tongue describe Her;
Yet, I want to earn Her favor

to bring Joy to my heart
and to renew my soul
so that it will be free from coarseness.

If my Lady grants me her Love,
I am ready to receive and reciprocate;
I can be either discreet or bombastic
to do and say whatever pleases Her;
to respect Her
and to praise Her.

I will attempt to speak to Her
although I am afraid to offend Her;
because I am scared to fail
in my declaration of love;
yet I know She will do what is best for me
because I shall be saved through Her.

As we can see, this song is not about ordinary sensual or romantic love. The song indicates the existence of something that could not be described by words. It is something that man longs for and is attracted to – without being able to define what that thing really is. It is a sort of fleeting feeling that man occasionally experiences. Some five hundred years later, the same idea was presented by Berowne, a character from Shakespeare's play entitled *Love's Labour's Lost*:

From women's eyes this doctrine I derive:
They sparkle still the right Promethean fire;
They are the books, the arts, the academes,
That show, contain and nourish all the world:
Else none at all in ought proves excellent.
                    (*Love's Labour's Lost*, IV.3)

It could be said that man, in his ordinary state, is separated from a higher state of being. Yet, in those fleeting moments, he recognizes that. These feelings are the first indicators of the possibility of accessing that state. The poets often compare the ordinary man to a sleeper who is "asleep" to that higher state. Awakening, therefore, is like gaining access to that other reality. The experience of this type of awakening ("opening of the door") is indicated in the following excerpt from a poem by Jalaluddin Rumi, a 13th century Persian poet:[7]

One went to the door of the Beloved and knocked.
A voice asked, "Who is there?"
He answered, "It is I."
The voice said, "There is no room for Me and Thee."
The door was shut.
After a year of solitude and deprivation he returned and knocked.
A voice from within asked, "Who is there?"
The man said, "It is Thee."
The door was opened for him.

In the language of the Troubadours' poetry, such a "sleeper" is presented as a lover who is separated from his beloved. In this context, "separation" is a technical term. It is equivalent to being "disentangled" from a more complete state of being. In other words, the *lady* of the Troubadours symbolizes a dynamic element that allows entangling the ordinary state of man to his higher level of being. The *lady* of the Troubadours symbolizes the means for realizing such an entanglement. She provides a bridge allowing one to cross from the ordinary physical world into the other one, i.e., the invisible one. This is why the *lady*'s

---

[7] *The Sufis*, Idries Shah, The Octagon Press, London, 1989, p. 317.

physical form was never described. Instead, her effect on the lover was the primary purpose of those love songs.

In applying the formula "As above so below," we can recognize that the process of entangling man to his higher state is reflected in the laws of quantum mechanics. Of course, the process discovered by physicists is a greatly simplified form of that which applies to the human mind. In the language of quantum mechanics, the term "higher state" corresponds to a state in which simple forms of matter, such as photons and electrons, seem to be able to gain a certain degree of awareness. As we will see, it took more than 900 years to understand the link between these two forms of entanglement.

The original formula of the Troubadours did not last very long. Quite soon, it was modified and presented in a new format. This new format became known as "courtly love." In this modified version, the *lady* of the Troubadours was substituted with ... a physical person. Of course, such a substitution obliterated its dynamic function. The overall concept became sterilized; it was gradually converted into a moral or dogmatic formula. In the context of the medieval "love courts," the lady - as a symbol - was substituted by a woman of higher status, usually the rich and powerful female head of a castle. The lover would try to make himself worthy of her by acting bravely and honorably. He would engage himself in whatever deeds she might desire or subject himself to a series of ordeals to prove his love and commitment.

The main architect of that modified form was Eleanor, Guillaume's granddaughter. Eleanor was the Duchess of Aquitaine. Later, she became Queen of France. By her second marriage to Henry, Duke of Normandy, she also became Queen of England. When Eleanor retired to Poitiers, she devoted her entire resources to promoting the development of the love courts. Poitiers became the academy of the courtly arts to which the nobility from far and near came for instruction. Several future kings and queens and many future dukes and duchesses were educated on Eleanor's campus and returned home to model their courts on hers. Thus, the courtly ideal and the lyrical love poetry were distributed over Europe – together with the ideas contained in these forms. This way, the concept of courtly love was transmitted from its local form into a pan-European one.[8]

But the modification of the original theme did not stop there. Its further deterioration affected the literary form, which became known as the corpus of King Arthur. The author of the Arthurian material was the French poet Chrétien de Troyes (1160-1191). Chrétien was a courtier at Poitiers and a protégé of Eleanor's daughter, Marie. Consequently, in King Arthur's tales, the lady went through further modification. Namely, the lady was transformed into Arthur's dutiful queen. As Arthur was a hero-king, the potential admirers of the lady had to juggle between their allegiance to their king and their desire for the lady. This led to the formation of that famous ménage à trois where the lover's love is for the married woman, while her husband does not seem to regard the suitor as an enemy or even a rival.

After Chrétien's death, the theme was taken up by other poets. At this point, further deterioration of the theme took place. Namely, a Christianizing process began. Now, the lady of the Troubadours was presented as the Virgin Mary. Gradually, the overall concept was transformed into Mariolatry, i.e., the worship of the Virgin Mary. With that substitution, the original

---

[8] A more detailed account of the role and activities of the love courts may be found in *The People of the Secret* by Ernest Scott (The Octagon Press, London, 1983, Chapter "Love Court, Troubadours and Round Tables").

impulse became completely erased. Clearly, it was time to renew the entire process.

In the late 14ᵗʰ century, a sign of such a renewal appeared. This time it took on the form of a chivalric romance written by an anonymous author.

The Green Knight

People can die of mere imagination.

(Geoffrey Chaucer)

As we have seen, the love poetry of the Troubadours was transferred to a new medium – the legend of King Arthur. As always, such a transfer led to the gradual corruption of the original concept. At one point, the corruption was such that the original idea was hardly recognized. At that point, *Sir Gawain and the Green Knight* was written. *Sir Gawain and the Green Knight* is a chivalric poem written by an anonymous author. Of the author of this poem, nothing is known. It has been estimated that the manuscript was written around the year 1400. This would indicate that the anonymous poet was a contemporary of Chaucer and his *The Canterbury Tales.*

Sir Gawain and the Green Knight is considered to be one of the greatest pieces of Middle English literature, i.e., the form used from the late 12th century to the 1470s. The manuscript was rediscovered only in 1839 in the library of Henry Savile of Bank in Yorkshire. Henry Savile lived at the time of Shakespeare, i.e., in the late 16th and early 17th century.

It looks like the poem's purpose was to indicate, or even slightly caricature, a concept which was publicized within the Arthurian literature at that time. The anonymous poet used the character of Sir Gawain to demonstrate how irrelevant or even ridiculous the overall concept of the Round Table and the love court had become. In this way, a door was opened for the introduction of an updated form of "entanglement."

The character of the Green Knight is taken from the legendary figure of Khidr, the Green One, who is often equated with St. George.

Khidr is reputed to be the mysterious guide who instructs humanity through those capable of getting into contact with him. He travels the earth in a variety of guises and by unknown means. Sometimes, he may seem to return good for evil or evil for good. But what he is really doing is known only to a fortunate few. An account of his actions is given in the Koran (Sura XVIII) in the story of Khidr and Moses. Here is one of the versions of that story.[9]

> Moses was traveling through a desert when he saw a man whom he recognized as Khidr, the Green One. Moses asked Khidr if he could accompany him on his journey. Khidr answered that he could, with the proviso that whatever Khidr did, no questions were to be asked about it.
> The pact was made, and the two walked on until they came to a wide river, which they could not cross without a boat. There was one, belonging to an old ferryman, and it was his only means of livelihood. Khidr agreed to take it over and to moor it safely on the far bank.
> As soon as the two were safely over the water, however, Khidr made a hole in the bottom of the vessel and half-sank it in full view of the boatman. The boatman, quite understandably, was weeping and wailing.
> Although he had enough perception - denied to most

---

[9] *Journey with a Sufi Master*, H.B.M. Dervish, The Octagon Press, London, 1987, p. 55.

people- to recognize Khidr, Moses was unable to understand how a spiritual being could thus repay good with evil. And he said so. Khidr, however, reminded Moses that he had promised to ask no questions.

Nothing much happened for some time, until they reached a village where they begged for a drink of water. Nobody would give them a drop; indeed, the inhabitants of the place abused them and shouted that they were to leave immediately, for no strangers were welcome there. They made their way to the outskirts of the place, and Khidr suddenly stopped by the crumbling wall of a mud hut. Asking Moses to help him, he collected clay and repaired the wall.

"Holy One," said Moses, "I know that one should do good to enemies, but surely there was no need to go to such extremes. Perhaps simply abstaining from reproaching them would have been enough."

Khidr merely reminded Moses of his undertaking not to ask questions.

When the two travelers reached another village, they saw some children playing in a field. Khidr crept up to one of the children, a small boy, seized him, and held him in such a way that the child died.

This was indeed the last straw for Moses.

"Great and Holy Khidr," he said. "I have heard that there is a Great Design, and that evil happens so that there shall be good; but I cannot endure seeing this happen, for experiencing something is not the same as thinking about it. To me, what you are doing is abnormal and forbidden. I must part company with you, unless you can explain yourself."

"I shall certainly tell you what I have been doing despite our pact," said Khidr. "But when I have told you, you must immediately leave me, for you have shown that you cannot endure the experiences which are those of the emissaries of the Hidden World."

"In any case," answered Moses, "I shall have to leave you; for all the upbringing which I have had was the dedication to making me a better person than a miscreant, a murderer and a returning evil for good, and it cries out against what you have done."

"Hear, then, Moses, good man though you are," said the Holy One, "that there is always a meaning in what happens, and that one part of the Great Design is not complete without the other parts.

"I am myself working in accordance with a Plan which you do not see. Even I have only a part of the plan in my mind, for God alone knows the complete one. But just as you have knowledge greater than an entirely ignorant person, so I, too, have knowledge greater than yours. This knowledge makes me do things and not do other things, and these actions appear incomprehensible to you, just as you may yourself appear baffling to the totally ignorant.

"I know, for example, that a tyrant king is on his way to confiscate all boats for transporting his army. If that boat which I damaged had been sound, the searchers would have taken it, and it would never have been returned to the ferryman. In his old age, he would have starved to death. Now the confiscators will think that the boat is useless and leave it there. Presently a certain carpenter will arrive, and he will repair the boat and take it back to the old man."

"And the wall, returning good for evil – was this merely a gesture, something to teach me with or something to acquire merit?" asked Moses, who now felt a little ashamed.

"The people living in that village were, as you will imagine, wicked, greedy, and cruel. There is a pot of gold in that wall, which the father of some orphans living there left for them. The wall was crumbling prematurely: the children are not old enough to take possession even of their ruined hut, let alone to guard their gold, their patrimony. We have mended the wall so that it will last until the exact moment when the children will be able to claim their heritage and to keep it."

Moses was impressed, and he began to feel that there was indeed something supremely important about Khidr's mission. But then the vision of the murder, the cold-blooded killing of a small child, swam before his eyes. Surely there could be no possible justification for such a thing?

"The boy was killed," said the Holy One, "just as people of all ages are killed daily by disease, accident, and murder; in this case, it was because this child was destined, had he lived, to grow up to be one of the greatest evildoers who ever lived. Millions would have died who had as much promise and who were equally loved, through the bloodlust and the horrors, unimagined, which he was going to perpetrate."

Moses now fell on his knees and cried: "Holy One, let me accompany you! Let me make amends for my ignorance and stupidity!"

But Khidr would not agree, and Moses stayed imprisoned within his own limited portion of the Great Design.

Most people think of mystics as people who follow a path towards their own salvation, as in the familiar Christian tradition, or as teachers of disciples in the Indian one. But the true mystics, in addition to these elements, strongly emphasize a worldly and cosmic role. They are widely believed to be involved in the mysterious past, present, and future of human progress on this planet.

So, let's return to *Sir Gawain and the Green Knight* and take a closer look at how the anonymous writer dealt with the Arthurian motif at the end of the 14th century.

***

The poem starts with a New Year's Eve celebration at King Arthur's court in Camelot.

> In accordance with the court's custom, Queen Guinevere presides over the festivities. Gawain, Arthur's nephew, sits next to her. When the dinner is about to be served, King Arthur insists on having some fun. He says he will not eat until he hears some stirring story, or someone challenges him to a jousting duel.
> At this very moment, a mysterious horse rider suddenly enters the hall. The stranger is enormous and imposing. He and his horse are all green. He holds a green branch of peace in one hand and a huge gold axe in the other. He has no other weapons, neither a helmet nor a shield. Without any greetings and without introducing himself, the green man demands to know who the governor of this gathering is. "For I would gladly like to talk to him," he adds. He looks at the courtiers, and rolls his eyes up and down, as if trying to find out who among them is the most renowned. A dead silence falls upon the assembly of stunned knights and ladies. They are all marveling at what all of this could mean. They are waiting for King Arthur to respond.

The Green Knight[10]

King Arthur salutes the stranger and says: "To this lodging, you are welcome! I am the governor of this household. My name is Arthur. I kindly ask you to tell us your wish so we may learn about your purpose for being here."

"My desire is not to spend too much time here," answers the stranger. And he continues, "Your castle and your knights are known as the best, the bravest, and most honorable in the entire world. Therefore, I want to offer a game to test the true valor of your assembly."

Arthur thinks that the green man seeks a sort of combat with his knights: "Sir, if you seek a battle, you shall find it here." But the green man clarifies that his wish is not for a fight. "Your knights are but beardless children; no man here could match me – they are too frail. I desire only a pastime, a Christmas game." He explains that he is looking for a brave knight who would be bold enough to accept an exchange of blows. "If anyone among your assembly is fierce enough, then I will give him my axe to strike me – provided that I will get to return the blow in exactly a year and a day from now."

The whole hall falls into a dead silence. They are all

10 https://villains.fandom.com/wiki/Green_Knight

shocked by the strange conditions of the green man's
game. When none of those in the hall dares to move, the
green man exclaims: 'What! Is this Arthur's house, whose
unmatched valor is rumored throughout the entire world?
Where now is your pride, bravery, and fierceness?" And
the green man bursts into laughter.

The green man's words anger King Arthur. "It is
madness that you ask for," he says. "And since it is just
folly that you seek, therefore you deserve to find it! So,
give me your axe, and I will give you the reward you are
asking for."

Arthur steps out, grabs the axe, and is ready to strike the
stranger. The green man is unfazed; he is no more
disturbed than if someone had served him a drink of
wine. He dismounts his horse, pulls down his coat, and
exposes his neck for Arthur's blow. At this moment,
Gawain comes forward and addresses Arthur: "I pray
that this match should be mine!"

Arthur grants Gawain's request, saying: "Take care,
cousin. Deliver one good blow to teach him a lesson. I
am sure you will be able to bear any blow he might give
back later."

The green man is visibly pleased with the course of
events. He says, "Sir Gawain, I am pleased to find that it
is you who will do me the favor that I ask."

At this point, we may realize that Gawain is the intended
target of the green man's challenge. Gawain is aware that
Arthur's rashness has led him into this risky situation. Gawain's
purpose, therefore, is to rescue his elder kinsman, his king, and
the head of the Round Table. Gawain's action shows that he is
the noblest and bravest member of the court.

Gawain walks towards the Green Man. The stranger asks
if he may trust that Gawain will respect the game's rules.
Gawain confirms that "at this time, a twelvemonth from
now, it will be your turn to deliver a blow with a weapon
of your choice."
The other man further demands that Gawain assures him
that he will search for him. Gawain asks, "Where can I
find you? I do not know where you live. Neither do I
know your name nor your castle." The green man
answers that he will tell him all that after taking Gawain's
blow. "Now," he says, "take the axe and strike me."
Gawain grips the axe, raises it, and swiftly drops it on the
stranger's neck. The stranger's head falls to the floor and
rolls around the hall as the knights fend it with their feet.
But, to everyone's amazement, the headless corpse
reaches down, picks up the head, and walks with it
toward the horse. He mounts the horse while keeping the
head by the hair in his hand. Then he turns the head
towards the terrified assembly. The head lifts its eyelids
and looks at them. Then the head starts to talk:
"Get ready, Gawain, to go and find me as you have
promised. I am known as the Knight of the Green
Chapel. You shall go to the Green Chapel, which is
known to many. You will find me waiting for you and
ready to return the blow there. If you fail to do so, you
will deserve to be called a coward."
Saying this, the Green Knight rides out of the hall with
his head in his hand.
Arthur and Gawain decide to hang the axe above the
main dais. King Arthur then addresses the queen, "Dear
Lady, be not downcast at all! Such cunning play well
becomes the Christmas tide, an interlude, just like the
singing and dancing of knights and dames. But now, to
my dinner, for I was granted a marvel that I wanted."
So, the entire assembly returns to their feast and

continues the festivities.

A year later, Gawain gets ready for his journey. It is time to depart in search of the Green Knight and the Green Chapel. And he knows that the only probable outcome of his journey – will be his death. He meticulously prepares his outfit.

The author of the poem describes quite precisely Gawain's outfit, particularly his shield. There are three entire stanzas dedicated to the shield's description. It is apparent that the design of the shield is an important element of the overall narrative. In his description, the poet introduced significant changes to the heraldic emblems used in the Arthurian romances. These changes and their disproportionately long and detailed description further underline the poet's overall intention. So, let's take a closer look at Gawain's shield.

Instead of the usual heraldic emblems of griffons or lions or eagles, the main motive on Gawain's shield is ... a pentangle, the Star of Solomon:

Sir Gawain's shield has a golden pentangle inscribed upon it. It is a sign of Solomon and an embodiment of Truth. It is a figure that has five points linked together by five lines. Each line overlaps and is linked with another, and in every way, it is endless. The English, I hear everywhere, name it the Endless Knot.

It was the first time that the term "pentangle" appeared in English literature. The original function of the five-pointed star was to be used as a symbol of the entanglement of the ordinary human mind with higher states of awareness.

According to a tradition, King Solomon (10th century BC) introduced a set of symbols to transmit knowledge about heightened states of human awareness. Like most symbols used in the transmission of knowledge, the Star of Solomon is not static; it is in permanent motion. This is why it is often called the "endless knot." It is "endless" because it replicates itself geometrically, i.e., every pentangle has a smaller pentagon that allows a pentangle to be embedded in it (see the following figure). This process may be repeated forever with pentangles of decreasing sizes. These imbedded stars are a representation of the various levels within the overall structure of the human mind. The important feature of this representation is that all levels are linked or entangled. Each level is determined by five characteristic features (points). This gradient-like design illustrates the overall structure of the mind.

A five-pointed star as applied to the human mind:
- External star: physical senses
- First inner star: subtle faculties
- Second inner star: the world of templates
- Third inner star: the world of ideas
- Inner dot: the Absolute

In the context of the development of higher levels of the human mind, the largest star corresponds to the five physical senses – which define the level of awareness of an ordinary man. The second star represents subtle faculties of the mind which, for an ordinary man, remain in their latent states. Specifically, the star signifies a template with five specific points on or around the human body. When this template is correctly fixed in mind and concentrated upon, these points provide an ambiance which allows the mind to resonate with higher levels of awareness. While in the resonance, the latent faculties become entangled with those higher levels, and, as a result, they can be activated. One of the initial marks of the activation of these faculties is a fleeting experience of a heightened state of mind. It is this fleeting experience that the Troubadours allegorically described as the unattainable *lady*. The *lady* was unattainable because initially, this state occurs transiently; it is not permanent yet.

The following stars of the pentangle correspond to higher levels of awareness that belong to the world of templates, the world of ideas, and the Absolute.

There are many other descriptions of such a structure of the human mind. For example, Shakespeare used the biblical story of Jacob and Laban to illustrate the entanglement between the various levels of the human mind. Shylock quotes the story in *The Merchant of Venice*. In this story, Jacob, inspired by an Angel, used partially peeled sticks that he placed in front of breeding ewes. As a result, the ewes gave birth to partly colored eanlings. According to his contract with Laban, all partly colored eanlings became Jacob's property. The Angel, Jacob, the conceiving ewes, and the eanlings represent the various levels of the human mind. The peeled sticks represent a template that is projected from the Absolute. The template contains the pattern of a more advanced structure of the mind. The template passes through the various strata of the human mind. Jacob represents a divinely inspired man whose role is to pass this pattern to ordinary men. In this way, the pattern may be entangled with the ordinary mind. The birth of partly colored eanlings marks the activation of new faculties of awareness.

Let us go back to Sir Gawain and the description of his shield. Obviously, the poem's author was familiar with the original meaning of the pentangle. Yet, to make his point, he presented it as a caricature of medieval religious beliefs that were prevalent at that time. He gives the following interpretation of the pentangle:

> The pentangle represents five sets of five Gawain's virtues.
> - his five senses;
> - his five fingers;
> - the five wounds Christ received on the cross;
> - five joys of Mary, Queen of Heaven,
> -the five knightly qualities: generosity, friendliness, chastity, chivalry, and piety.
> All these five sets of virtues are Gawain's trademarks, and in no circumstances is he to stain them.

For those familiar with the pentacle's original meaning, the above description is a ridiculous mix of non-related elements. By skillfully caricaturing it, the poet indicates how some developmental tools and concepts had become corrupted and misused at that time. Now, let's continue with Gawain's adventures.

> Gawain heads out into the wilderness in search of the mysterious Green Knight. He encounters various foes – wolves, bulls, bears, boars, and a variety of wood-trolls and other monsters. As the weather becomes colder, he nearly freezes to death. Thus, in peril and pain, he keeps searching till the day of Christmas Eve. On that day,

desperate as he is, he prays that he may find a place to
attend Christmas Mass. It is then that a beautiful castle
appears in front of his eyes. Surrounded by a pleasant
park and moat, the castle shines through an oak forest in
the distance.

It may help to grasp the meaning of the sudden appearance
of a castle to recall the following story.[11]

A knight of chivalry got lost in the wilderness. Suddenly,
he saw a magnificent castle appearing from nowhere in
front of his eyes. He stopped, perplexed. A steward came
out of the castle and said:
"My master, the owner of this castle, invites you to enter.
There are refreshments and diversions if you would care
to consider yourself our guest."
The castle was full of servants. The knight was
overwhelmed with its splendor and luxury. He was led to
a banqueting hall where every kind of delicious food was
waiting for him. When the meal was finished, the host
took him to see his gardens, which covered an immense
extent of land. There, amid every conceivable variety of
fruits and flowers, an army of gardeners was at work,
swarming like ants through the grounds. All day, the
knight kept asking his host to explain to him what all of
this meant. But the host would only say to all his
entreaties:
"Wait until morning."
Morning came, and instead of waking in the luxurious
silken bed to which he had been conducted the night
before, the knight found himself lying stiff and stark on

[11] Adapted from *The Magic Monastery* by I. Shah, The Octagon Press, London,
1984, p. 13.

the ground within the stony confines of a huge and ugly ruin on a barren mountainside. There was no sign of the host, of the beautiful arabesques, the libraries, the fountains, the carpets.

"The infamous wretch has tricked me with deceits of sorcery!" shouted the knight.

But what he did not know was that, by the same means that he had used to conjure up the experience of the castle, the host had made him believe that he was abandoned in a ruin. He was, in fact, in neither place. The host now approached the knight, as if from nowhere, and said:

"We shall now return to the castle."

He waved his hands, and the knight found himself back in the palatial halls.

The host waved his hands again, and the knight found himself in the wilderness, exactly in the same spot where he was when the castle had first appeared to him. In fact, he had never left that spot. Then he heard a voice saying:

"As long as you are driven by greed for knightly fame, respect, and recognition – so long it is impossible for you to tell self-deceit from reality. There is nothing real that can be shown to you – only deceit. Those whose food is self-deceit and imagination – can be fed only with deception and imagination."

In other words, the appearance of such a castle indicates that the hero, in this case Sir Gawain, is being led through a series of experiences that will allow him to discover the truth about himself. So, let's see what kind of tests have been prepared for him in the castle that appeared to him on that day.

Gawain approaches the drawbridge. He calls, and a porter appears. "Good sir," says Gawain, "will you go with my message to the lord of this house. I seek a place where I could say my prayers." The porter leaves and soon comes back and welcomes him to enter.

The lord of the castle comes to meet him and says, "You are welcome to dwell here. What is here, all is your own." The host is quite an imposing figure. Though his powerful appearance makes him look fierce, his manners are courtly. The host assigns servants who prepare Gawain's chamber. Later, the host introduces Gawain to his wife. The host's wife is young, beautiful, and elegantly dressed. Gawain realizes that the host's wife is more beautiful than Queen Guinevere. During the conversation, the host enquires about Gawain's purpose for traveling. He and his court are pleased to learn that their guest is a knight of Arthur's Round Table. After the meal, the men and ladies play games and celebrate late in the night when Gawain retires for bed.

Gawain spends Christmas Day and the two days following in a similar fashion. With only three days remaining before his appointed meeting with the Green Knight, Gawain asks for permission to leave. He explains that he must search for the Green Chapel. The host laughs and replies that there is no problem with Gawain's quest – for the Green Chapel is only two miles away. Delighted, Gawain agrees to stay three days longer - until New Year's Day. The host tells Gawain that he plans to spend three days hunting. And he proposes a game. During the day, he wants Gawain to stay at court and enjoy himself in the host's wife's company. At the end of each of these three days, the two men will exchange their gains, i.e., whatever the host catches in the wood shall be Gawain's, and whatever Gawain gains during his stay at the castle - he shall give to his host. The host describes

his arrangement with Gawain as a "game," suggesting
that the challenge is no different from any of the other
games played by Arthur's court. At the same time,
however, the host refers to the agreement as a pact, a
contract, making sure that Gawain understands its
provisos. Happy to play along, Gawain vows to comply:
"while I remain in your castle, your command I'll obey."
The men kiss each other and retire to their chambers.
Early in the morning, the host and his men leave to hunt
deer with their hunting dogs. Respecting the hunting
season rules, they do not kill the male deer. Instead, they
separate the does away from the bucks and harts, then
wound them with arrows. Afterwards, the dogs hunt
down the wounded animals, and only then do the hunters
kill them off with their knives.
Back at the castle, Gawain stays in bed until daybreak.
While still half asleep, he hears that his chamber door
opens quietly. Peeking out of his bed's curtains, he sees
the host's wife walk into the chamber and cautiously
close the door behind her. Gawain pretends to be asleep.
The lady casts aside the curtains, climbs inside the bed,
and sits beside Gawain. Gawain is not sure what to do.
He rolls over, opens his eyes, and pretends he is
surprised. He makes the sign of the cross, which makes
her laugh. She greets him and jokes that he is a careless
sleeper because he can easily be captured. Gawain laughs
and says that, in such a case, he is willing to surrender to
her. Then, laughingly, he asks her to release him so he
can get up and put on his attire. The lady refuses, saying
that she will hold him captive because it is a rare privilege
to spend some time alone with a famous and honorable
knight. She offers to be his servant and tells him to use
her any way he wishes.
Gawain responds that serving such a gracious lady would
be an honor. The lady persists in lavishing Gawain with

compliments. The two continue their courteous bantering till mid-morning. Then the lady gets up, and while leaving, she accuses Gawain of not being a true knight. Surprised and worried that he has failed in his courtesy, Gawain asks for an explanation. The lady responds that a true chivalric knight would never let a lady leave his chamber without taking a kiss. "Very well," says Gawain, "as you wish, I will kiss you at your command." The lady takes him in her arms and kisses him. Then she withdraws and departs. Gawain dresses up and goes to hear mass. He spends the afternoon in the company of the host's wife and other ladies.

In her bartering with Gawain, the lady evokes the code of courtly love and chivalry. According to this code, a knight's love for a lady of higher rank leads to his spiritual ennoblement. Therefore, Gawain is required to obey her. However, to accept her advances, he would have to break his chivalric code of chastity. By invoking these rules during their erotically charged moments, the lady puts Gawain in an impossible situation. It is obvious that the poet is emphasizing the artificiality of the courtly love that had been popularized in the Arthurian literature.

In the evening, the host returns to the castle. He greets Gawain and gives him the venison he caught during the day. Gawain accepts the venison and, in return, kisses him and says: "There take you my gains, sir! I got nothing more." The host laughs and asks: "Tell me where did you gain this kiss?" Gawain replies that they agreed to exchange their gains but not to tell how they acquired them. The men decide to continue the game the next day. On the second day, the host hunts a wild boar. This time

he has to wrestle the beast to the ground and then kill it
with his sword. At the castle, the lady continues to tease
Gawain by arguing that Gawain's acceptance of her love
agrees with the chivalric code. Again, she challenges his
knightly reputation by demanding two kisses. Gawain
answers, "I am at your call and command to kiss when
you please. You may receive as you desire." The lady
kisses him and leaves the room.

That night, the host and his men return to the castle,
carrying the boar's head. The host gives Gawain his
catch, and in return, Gawain gives him two kisses.

On the third day, the host hunts a fox. At the castle,
Gawain is again awakened by the lady. This time he gets
three kisses from the lady. While they keep bantering, the
lady asks Gawain for a love token: "Now, at this parting
do me this pleasure and give me something as your gift
that I may remember you." Gawain refuses her request,
claiming he has nothing to give: "I am here on an errand
in unknown lands, and have nothing to give." Therefore,
the lady offers him a gold ring with a red stone. Gawain
again declines to accept it: "I will have no gifts at this
time. I have none to return, and naught will I take."

The lady's approach parallels that of the fox chase. Like
the hunter, she uses more unpredictable challenges than
during the previous two attempts. Namely, she offers
Gawain her green belt, which she claims has magical
properties: "For whoever carries this belt cannot be killed
by any cunning of hand." Gawain, like the fox, fears for
his life and is looking for a way to avoid death from the
Green Knight's axe. Tempted by the possibility of
protecting his life in the upcoming appointment with the
Green Knight, Gawain accepts the belt. However, he
must promise her that he will not tell the host about it.

That afternoon, Gawain goes to confession. After the
confession, he is relieved and happy. Now he feels that

he is ready for the encounter with the Green Knight. He spends the rest of the day dancing and enjoying the lady's company. Yet, when the host returns to the castle in the evening, Gawain gives him the three kisses, but he does not mention the lady's belt.

Confession is another important element of the poem. The poet seems to emphasize that Gawain is convinced that he truthfully confessed all his misdeeds. At this point, a question may be asked: did Gawain really confess all his offences? Afterall, he went to his confession prior to his final exchange of "gains" with the host, i.e., before he would commit his greatest offense: breaking his pact with his host. The poet seems to point out that it does really matter what Gawain confessed or what he did not. Instead, as we will see later, the poet uses the poem to illustrate the artificiality of the commonly accepted function of confession.

After the exchange, the host and his courtiers hold a farewell party for Gawain. Late at night, Gawain retires to his chamber to prepare himself for the journey to the Green Chapel. Before the sun comes up, he rises and puts on his armor. He remembers to tie the lady's belt around his waist.

As Gawain prepares to ride off, he wishes joy and happiness to the host and his wife. Accompanied by a guide, Gawain crosses the drawbridge and rides back out into the wilderness, up to the heights of the neighboring snowy hills. The guide tells Gawain that no one can survive an encounter with the Green Knight. The guide suggests that Gawain should abandon his quest and leave without facing the mysterious knight. The guide promises not to tell anyone: "I shall haste home, and on my honor,

I promise that I will keep safe your secret and say not a word that ever you were willing to flee."

Gawain thanks the guide for his concern but he refuses to accept his advice. The guide wishes Gawain well, shows him the way to the Green Chapel, and quickly leaves as he is afraid to go any farther.

Gawain heads towards the hills. He sees no sign of buildings or chapel. Then, finally, he notices a strange large barrow – like that used to cover the remains of the dead. In it, he spots an old cavern. He realizes that the cavern within the barrow must be the Green Chapel.

Suddenly, Gawain hears the horrifying sound of a blade being sharpened on a grindstone. Terrified and fully aware that the sound means his death, Gawain announces his arrival: "Who is master in this place to meet me at this rendezvous?" A voice replies from the top of the cavern, telling Gawain to stay put. Gawain hears that the man continues to sharpen his weapon. After a while, the Green Knight emerges carrying an axe. He welcomes Gawain warmly and compliments him on his punctuality: "I welcome you to my place; you have timed your travels as a trusty man should, and you have not forgotten the engagement agreed upon between us."

The Green Knight tells him to prepare to receive the blow he was promised. Gawain tries to act unafraid as he bares his neck and gets ready for the deadly blow.

The Green Knight lifts the axe high and drops it. When the Green Knight sees Gawain flinch – he stops his axe, and he mocks Gawain for his cowardice. Gawain promises not to flinch again. This time Gawain doesn't flinch as the axe comes down - but the Green Knight holds the axe again. He congratulates Gawain for his courage. He then says that the next blow will strike him. Angry, Gawain tells the knight to hurry up and strike. The knight lifts his axe one more time and finally strikes

Gawain but gives him only a small cut on the neck. Gawain springs up and grabs his sword, telling the knight he will defend himself now that he has taken the promised blow. The Knight leans on his axe and agrees that Gawain has met the covenant terms but refuses to fight. He explains that he feinted the first two times because Gawain respected the contract they made in the castle. Now Gawain realizes that the Green Knight is the host of the castle where he was staying. The Green Knight reminds Gawain that he was truthful to their pact only on the first two days. However, the small cut from the third blow was punishment for Gawain's misdeed on the third day. On that day, Gawain did not tell the truth about the green belt. Therefore, he did not pass the loyalty test; he betrayed his host.

Gawain responds by untying the belt, cursing it, and giving it to the Green Knight. He asks to regain the host's trust. "I am cursed, coveting, false, and a coward," he shouts. "Now I confess *truly*, sir, here to you all my faults!"

It is only at this point that Gawain is capable of adequately confronting and recognizing his misdeed. The circumstances were such that he was able to *experience* a true confession.

The Green Knight laughingly accepts Gawain's confession and offers him the green belt as a souvenir. He asks Gawain to come back to the castle and stay there longer to celebrate New Year's with him, the lady, and his knights. Gawain refuses the invitation.

However, he accepts the belt. He puts the green belt on his right shoulder and hurries back toward Camelot. When he enters the court in Camelot, he meets a delighted reception. The whole court rejoices to see him safe and hear his marvelous tale. Gawain explains that he will always wear the belt to remind him of his fault. The king and knights laugh about it and decide they will also wear green belts for Gawain's sake.[12]

By refusing the invitation of the Green Knight, Gawain fails his test. Just like Moses in his encounter with Khidr, Gawain demonstrates that he is incapable of benefiting from the company of a wise man. At this point, it may seem that the effect of his encounter with the Green Knight in the Green Chapel did not cause any change in his inner state. But .... the story is not finished yet. We will meet Gawain again in the next chapter.

Now we can see what the overall purpose of the poem was. It was to indicate the irrelevance of the chivalric concept that was being popularized in Arthurian literature at that time. At the same time, the poet indicates that there is another way to develop and exercise truly chivalric virtues. In other words, it was time for the renewal and reactivation of the developmental process that the Troubadours introduced. Indeed, the appearance of *Sir Gawain and the Green Knight* corresponded to the time when a revival of the process was initiated.

---

[12] Adapted from the translation by J.R.R. Tolkien, *Sir Gawain and The Green Knight with Pearl and Sir Orfeo*, Harper Collins Publishers Ltd., London, 2021.

Epilogue
to *Sir Gawain and the Green Knight*:

A man of aristocratic look walks into the reception hall of
a modest castle. A servant ushers him to a room where
the host rises from a chair and welcomes him warmly.
After exchanging the customary greetings, the visitor
takes a large manuscript out of his bag and hands it to the
host. The host asks him to take a seat and then opens the
manuscript. On the title page is written *Sir Gawain and the
Green Knight*. After turning a few pages, the host takes a
pen and writes on the last page of the manuscript:

HONI SOIT QUI MAL Y PENCE

Then, he returns the manuscript to the visitor, saying:
"Well done, my friend! This part of your journey has
been completed. You have succeeded in outgrowing your
boastful Gawain and your impatient Moses. You have
arrived at the second Green Chapel. Now, we must part
again. But this time, you know what is ahead of you; you
know your way. Look for a … *pearl*.
So, farewell, and have a star to guide you."
The visitor kisses the hand of the host and leaves.

***

The appearance of the phrase HONI SOIT QUI MAL Y
PENCE in the manuscript of *Sir Gawain and the Green Knight* is a
hint about the overall purpose of the poem. The origin of this
phrase can be traced to the formation of the Order of the
Garter, which was created by King Edward III around the year
1348. King Edward's aim was to revive the tradition of the
Round Table.

The Order of the Garter was structured similarly to the Khidr Order – which was founded in the East a century and a half earlier. (The Khidr Order was also known as the Order of the Round Building – after the great palace of Bagdad that belonged to Haroun el-Rashid.[13]) The Khidr Order was composed of groups called "circles." Each circle consisted of thirteen members. Their purpose was to develop an understanding of what lies behind the limitations of ordinary awareness. As their slogan, they took an Arabic motto about Saki, i.e., a cupbearer, their guide.

According to a mystical legend, Saki was to serve God every day a cup of wine. Every morning, the cup would be replenished with wine only if Saki had discharged his daily responsibilities and duties correctly. On the fortieth morning of his impeccable service, Saki was presented with a cup of the Drink of the Immortals. This legend inspired many tales about the challenges a hero must undertake to fulfill his heart's desire. For example, here is Omar Khayaam's reference to Saki:

> Each drop of wine that Saki negligently
> Spills on the ground may quench the fires of grief
> In some sore heart. All praise to Him who offers
> Such medicine to relieve its melancholy![14]

Like the Khidr Order, the Order of the Garter was also divided into circles (garters) of thirteen members each. One circle was presided by King Edward III, and the other circle was under the direction of the Black Prince, Edward III's son. At the same time, King Edward began to build a gigantic circular

13 *The Sufis*, Idries Shah, The Octagon Press, London, 1989, p. 220.
14 *A Journey with Omar Khayaam*, W. Jamroz, Troubadour Publications, Montreal, 2018.

building within the upper ward of his castle to house the Order there.

The phrase "Honi soit qui mal y pense" was chosen as the slogan of the Order of the Garter. Phonetically, this phase sounds like the Khidr Order's members' salutation to Saki. (Later on, the meaning of this phrase was explained as "Shame unto him who thinks evil of it;" this was the closest possible translation of the sound of this phrase into Anglo-Norman-French.) This slogan marked a major adjustment in the process that was implemented within Western society at that time. It was not a king who presided over the process. Instead, it was the Green One, a guide, who guided the Order. Another interesting thing about the Order of the Garter was a restatement of the original idea of the lady of the Troubadours. Namely, despite the exclusively male nature of the Order, Edward III included a woman to be associated with each of the original circles, i.e., Queen Philippa and Isabella, Edward's daughter. These two ladies were issued with robes and hoods. They were participating in the annual feast of the Order of the Garter, which was held at Windsor Castle.

All of these indicate that the appearance of *Sir Gawain and the Green Knight* marked the renewal of the process that the Troubadours previously initiated.

The Child of the Heart

The man who knows must be aware that
the child of the spirit is born in one's heart.
(Qadir Jilani)

Gawain's encounter with the Green Knight had a lasting effect on him. Although at first, he refused the Green Knight's invitation to join him and his knights, it looks like after some time, he returned to the Green Knight's castle. It was then that Gawain went through a series of experiences which are described in the poem entitled *Pearl*. *Pearl* is a sequel to *Sir Gawain and the Green Knight*. We may realize that the poet used the character of Gawain to describe his own experiences. Both poems were contained in the same manuscript. It is believed that they were written by the same person who became known as the Pearl Poet.

If *Sir Gawain and the Green Knight* was an announcement of the renewal of the process initiated by the Troubadours, then *Pearl* disclosed the updated template of the human mind which was projected at that time. *Pearl* gave insights into the next level of the structure of the human mind. And this design also provides further hints about the overall structure of matter which is of interest to us.

The title of poem, *Pearl*, refers to a symbolic character whose function corresponds to that of the *lady* of the Troubadours. The songs of the Troubadours described a certain longing, a powerful desire, an aspiration towards an unknown but strongly felt attraction symbolically presented as a *lady*. The songs implied that the *lady* was unattainable, and her suitor's love was hopeless. The interesting thing was that the songs described the *lady*'s impact on her suitor. The songs, however, did not describe how this kind of feeling could be fulfilled or how it could be developed further. They were focused on bringing attention to certain feelings, which were the very first indication of something much deeper and stronger than usual emotional or intellectual stimulations. In other words, the purpose of the Troubadours' songs and poetry was to make man aware of such feelings. However, the Troubadours did not address the next stages of the process. It looks like this task was left for somebody else. And it is this void that the anonymous poet is fulfilling in *Pearl*. The character of Pearl illustrates further the function of the *lady*. As pointed out in the previous chapter, all these activities were related to the formation of the Order of the Garter. It looks like the Order of the Garter was formed to advance the process to its next stage.

***

Familiarizing ourselves with some of the technical terms which had been introduced in the literature previously, i.e., prior to the late 14th century, will help to grasp the meaning of *Pearl*. Namely, an elevated state of awareness is often referred to as a "child of the heart." In this convention, this *child* is within the human mind (often referred to as "heart"). It is also within the mind that this *child* is fed and grows. The purity of the child is often represented by various forms of physical beauty. Sometimes, it is represented in the form of angels. Through this child, one can gain access to higher levels within the structure of the human mind. Such access allows one to enhance one's perceptiveness. For example, Qadir Jilani, a Persian mystic who lived in the 13th century, wrote in his *The Secret of Secrets*:[15]

> The man who knows must be aware that
> the child of the spirit is born in one's heart.
> This is the sense of true humanity:
> You must educate the child of the heart,
> teaching oneness by constant awareness of oneness,
> leaving this world of matter and multiplicity,
> looking for the spiritual world, the world of mysteries,
> where there is nothing but the Essence of God.
> There's really no other place but that place
> which has no beginning and no end.
> The child of the heart flies over that infinite field
> seeing things no one has ever seen before
> that no one could tell or describe.
> Just as Jesus said:
> "Man must be born twice to reach the kingdom of angels,
> like birds, who are born twice."
> That possibility lies in man.
> That is the mystery, the secret of man.

Often, this *child* is referred to as a "hidden pearl." A pearl is a symbol of enlightenment, wisdom, spiritual energy, and hidden beauty. As a feminine symbol, it comes from the forming of the pearl inside an oyster. *Pearl*, therefore, is a symbolic reference to a seed that was cultivated within the minds of 14th century Western society.

---

[15] *The Secret of Secrets*, Abdul-Qadir al-Jilani, translated by Tosun Bayrak (The Islamic Text Society, Cambridge, U.K., 1992).

***

The process described in *Pearl* takes place within a multiple-level structure. The earthly environment is the lower, ordinary level. The higher levels correspond to the invisible worlds. These higher levels are often referred to as divine or heavenly. In this structure, the mind of an ordinary man is seen as being separated from those higher levels. The purpose of the overall process is to activate certain faculties which allow the mind to operate in these higher realms.

Now, let's go to our poet and follow him through the series of experiences he was exposed to after his initial training at the Green Knight's castle. After rejoining the Green Knight, Gawain was given some exercises which would help him to gain access to higher levels of awareness. It looks like a *pearl* was an object he was to meditate on. The *pearl*'s image induced in him an elevated state of completeness and serenity. The poet adds, "two years you lived not on earth with me." In this way, he indicates that he has "cultivated" this experience for nearly two years.[16]

At the poem's beginning, the poet is distraught at losing his *pearl*. He describes the *pearl* as priceless and of unmatched beauty:[17]

> Never proved I any in price her peer.
> So round, so radiant ranged by these,
> So fine, so smooth did her sides appear

---

[16] Coincidently, it usually takes about two years to form a pearl. Natural pearls are formed when the oyster reacts to an irritant – by coating it with the shiny iridescent material found on the inner surface of the shell.

[17] The quoted verses are from the translation by J.R.R. Tolkien, *Sir Gawain and The Green Knight with Pearl and Sir Orfeo*, Harper Collins Publishers Ltd., London, 2021.

That ever in judging gems that please
Her only alone I deemed so dear.

It looks like the poet could not hold onto his elevated state. He describes this interruption as the *pearl* escaping from him ("from me it shot"/ "it sped from me") and disappearing in a spot on the ground of his garden:

Alas! I lost her in garden near:
Through grass to the ground from me it shot;
I pine now oppressed by love-wound drear
For that pearl, mine own, without a spot.

Following the Troubadours' code, he assigns to the *pearl* beloved-like attributes which uplift and heal. When the feeling of completeness is gone, the poet is distraught and sad:

Since in that spot it sped from me,
I have looked and longed for that precious thing
That me once was wont from woe to free,
to uplift my lot and healing bring,
But my heart doth hurt now cruelly,
My breast with burning torment sting.

In the same stanza, the poet discloses another essential detail. Namely, he indicates that he "cultivated" his *pearl* during his meditation, to which he refers as the "secret hour." This means that during nearly two years, he developed his mind to a state which allowed him to establish a link with an extraordinary source of beauty, fulfillment, and joy.

After the loss, he continued his meditations. And it was during one of these meditations that he was provided with further guidance in the form of "the sweetest song":

> Yes in secret hour came soft to me
> The sweetest song I e'er heard sing;

Shakespeare describes a similar experience in his *Twelfth Night*. Duke Orsino is referring to hearing and then losing such a "sweet sound":

> If music be the food of love, play on;
> Give me excess of it, that, surfeiting,
> The appetite may sicken, and so die.
> That strain again! it had a dying fall:
> O, it came o'er my ear like the sweet sound,
> That breathes upon a bank of violets,
> Stealing and giving odour! Enough; no more:
> 'Tis not so sweet now as it was before.
> (*Twelfth Night, I.1*)

By following the *song*, the poet finds himself in an unknown world. While wandering there, he comes across a stream he cannot cross. This experience marks the first signs of the activation of some of the subtle faculties. Now the poet can "see" beyond his ordinary senses.

Pearl – illustration by Simon Armitage (*The London Magazine*)

Then, on the other side of the stream, he sees a young maid.
It is only at this point that the poet perceives the manifestation
of his previously experienced longing in the form of a young
girl:

> A child abode there at its base:
> She wore a gown of glistering white,
> A gentle maid of courtly grace;
> Erewhile I had known her well by sight.
> As shredded gold that glistered bright
> She shone in beauty upon the shore.

The longer he is looking at her, the more certain he is – that
the maiden embodies his lost *pearl*.

> Long did my glance on her alight,
> And the longer I looked I knew her more.

The poet is asking the maiden whether she is the *pearl* whom he has lost:

> "O Pearl?" said I, "in pearls arrayed,
> Are you my pearl whose loss I mourn?"

The maiden starts to talk to the poet. Now, we may learn a bit more about the developmental function of the *lady* of the Troubadours. We may find out what the purpose of that experience was.

The maiden corrects the poet by telling him that he has lost nothing. She compares the poet to a jeweler and herself to a *pearl* that has been protected in a safe casket:

> Good sir, you have your speech mis-spent
> To say your pearl is all away
> That is in chest so choicely pent,
> Even in this gracious garden gay,
> Here always to linger and to play
> Where regret nor grief e'er trouble her.
> 'Here is a casket safe' you would say,
> If you were a gentle jeweller.

She explains that she is in a place that she really belongs to. Therefore, the poet's grief is completely misplaced:

> You grudge the healing of your grief,
> You are no grateful jeweller.

This means that the poet is exposed to the next stage of his experiences. His previous experiences served as preparation. Now, he faces a much more challenging task.

The poet apologizes and wants to cross the spring to be closer to her. Her reaction is even more unexpected. She scorns him by implying that either he is jesting, or he is mad:

> Why jest ye men? How mad ye be!

She says that the poet's reaction is thoughtless and witless. There is no way that he could cross the stream to join her:

> You believe I live here on this green,
> Because you can with your eyes me see;
> Again, you will in this land with me
> Here dwell yourself, you now aver;
> And thirdly, pass this water free:
> That may no joyful jeweller.

And she adds that he would need to take another course of action to be able to cross the water:

> Now over this water you wish to fare:
> By another course you must that attain.

The poet does not understand and starts to complain:

Why must I at once both part and meet?

The lady starts to lecture him. She tells him that by lamenting one thing, he may forsake something much more valuable:

Through loud lament when they lose less
Oft many men the more forgo.

The maiden continues her tutoring. She tells him that her current position within the invisible hierarchy is that of a queen:

A blissful life you say is mine;
You wish to know in what degree.
Your pearl you know you did resign
When in young and tender years was she;
Yet my Lord, the Lamb, through power divine
Myself He chose His bribe to be,
And crowned me queen in bliss to shine,
While days shall endure eternally.

The fact that the maiden is a queen within the invisible hierarchy shocks the poet:

"O Blissful!" said I, "can this be true?
Be not displeased if in speech I err!
Are you the queen of heavens blue,
Whom all must honour on earth that fare?
We believe that our Grace of Mary grew,
Who in virgin-bloom a babe did bear;
And claim her crown: who could this do
But once that surpassed her in favour fair?"

The poet doubts that the maiden, being so young, could be elevated to such a high state in the heavens. He is wondering whether the maiden has replaced Mary as Queen of the heavens:

You in heavens too high yourself conceive
To make you a queen who were so young.

The maiden explains that every being who can cross the water becomes queen or king in this place:

The court where the living God doth reign
Hath a virtue of its own being,
That each who may thereto attain
Of all the realm is queen or king,
Yet never shall other's right obtain,
But in other's good each glorying
And wishing each crown worth five again,
If amended might be so fair a thing.

In the same stanza, the maiden adds another detail about the structure of the invisible word:

But my Lady of whom did Jesus spring,
O'er us high she holds her empery,
And none that grieves of our following,
For she is the Queen of Courtesy.

The maiden refers to the structure of the invisible worlds. She says that the Queen of Courtesy belongs to a higher level within the invisible structure. The first level is that of the world of templates; the next higher one is called the world of ideas. "Courtesy" is one of the attributes that is associated with the world of ideas. What is important here − is that the world of ideas is beyond the reach of any earthly mind. That level is perceptible only by the minds who are already within the world of templates. This applies to the minds that managed to disconnect themselves completely from earthly influences. By placing the Queen of Courtesy within the world of ideas, the maiden emphasizes an important detail: concentrating one's attention on the Queen of Courtesy is useless for earthlings. First, one would have to gain access to the level of the maiden. Only then the perception corresponding to the level of the Queen of Courtesy might be attained. (Although the poem is wrapped in religious terminology, we may notice that the meaning of those terms does not correspond to that commonly used.)

The maiden quotes Saint Paul's analogy to explain further the overall structure of the invisible worlds:

As head, arm, leg, and navel small
To their body doth loyalty true unite,
So as limbs to their Master mystical
All Christian souls belong by right.

Familiarity with the process and the overall cosmic structure may help to grasp the meaning of the maiden's explanation. Namely, the two-level structure of the invisible worlds is not static; it is like a growing organism. The growth is sustained by the minds that are able to overcome the limitations of earthly conditions. In this way, they become a part of a new being that is gradually formed within the invisible worlds.[18] Like any other being, this new body consists of numerous parts which are needed to make it fully functional. It is in this context that it may be said that even the smallest part of the new being must be perfect ("is queen or king") in discharging its heavenly function.

The lady continues her explanation:

"Thus I," said Christ, "shall the order shift:
The last shall come first to take his due,
And the first come last, be he never so swift;
For many are called, but the favourites few."

At this point, the maiden quotes Matthew's parable of the vineyard workers. A vineyard landowner goes out early in the morning and hires men, agreeing to pay them a penny a day. Later, he goes to the market and hires another group of workers. He hires more workers at various times throughout the day and promises them fair wages. When the end of the day comes, the landowner asks his manager to pay the workers. And he requests that all workers should be paid the same wage, i.e., a penny, and those who came last should be paid first. When it comes to those who have been hired first (early in the morning), they think they will receive more. They begin to wrangle when

---

[18] The process of the formation of this new cosmic being is described in *The New Cosmos* by W. Jamroz, Troubadour Publications, Montreal, 2021.

they, too, are given a penny. They are angry because, in their view, they have done a lot more work than those who started later in the day. The landowner does not listen to their complaints and reminds them that they have agreed to the daily pay rate when they were hired.

Again, it will help to look at Matthew's parable in the context of the new cosmic structure that is being gradually developed. The growth of the new cosmic structure is not linear. This means that later phases of growth require qualitatively greater skills and efforts. Secondly, certain elements of the new structure must be completed in a certain order and at specific times. Therefore, those "minds" that join the effort at the later stages of the process are required to make a greater qualitative effort. In this context, the meaning of "the last shall come first to take his due, and the first come last" may be grasped. Yet, all contributors will evenly benefit from the overall enterprise because they become part of a new advanced structure.

Now, let's go back to our poet. When the poet addresses the lady as "a matchless maid and immaculate," the maid refutes him, saying:

"Immaculate, without a stain,
Flawless I am," said that fair queen;
"And that I may with grace maintain,
But 'matchless' I said not nor do mean.
As brides of the Lamb in bliss we reign,
Twelve times twelve thousand strong, I ween
…
In the city of New Jerusalem."

These "twelve times twelve thousand" refers to what may be called the potentiality of the invisible worlds. This potentiality is symbolically described as "Twelve times twelve thousand strong." These are spots to be filled in with earthly "lovers." The maiden refers to this new structure as "the city of New Jerusalem." The maiden describes New Jerusalem in the following way:

> In the other naught is found but peace
> That shall last for ever without impair.
> To that high city we swiftly fare
> As soon as our flesh is laid to rot;
> Ever grow shall the bliss and glory there
> For the host within that hath no spot.

The maiden tells the poet that "New Jerusalem" is presided over by "the host," the perfect man who "hath no spot."

The term "New Jerusalem" is equivalent to the "New Cosmos." The term New Cosmos is used to describe an invisible cosmic structure that is formed from perfected minds. Its growth is the ultimate purpose of mankind.

The maiden refuses when the poet requests that she leads him to New Jerusalem. She tells him that he does not have enough strength to withstand the impact of that place; he is not ready yet for this experience:

> "God will forbid that," she replied,
> "His tower to enter you may not dare.
> For a sight thereof by favour rare:
> From without on that precinct pure to stare,
> But foot within to venture not;

In the street you have no strength to fare,
Unless clean you be without a spot."

The maiden informs the poet that he has been granted the rare privilege of being able to see the city from the outside. She tells him how he can get to the place from which he would be able to see New Jerusalem. The poet wanders for a while along the bank of the stream. Finally, when he is able to see it, he is struck by its beauty and magnificence. It is then that he sees a procession of crowned queens led by a Prince, the host of the place:

The best was He, blithest, most dear to prize
Of whom I e'er heard tales of yore;
So wonderful white was all His guise,
So noble Himself He so meekly bore.

He recognizes the maiden among those taking part in the procession. Despite the warning of the maiden, he cannot restrain himself from attempting to cross the water that separates him from her:

When I saw her beauty I would be near,
Though beyond the stream she was retained.
I thought that naught could interfere,
Could strike me back to halt constrained,
From plunge in stream would none me steer,
Though I died ere I swam o'er what remained.
But as wild in the water to start I strained,
On my intent did quaking seize;
From that aim recalled I was detained:
It was not as my Prince did please.

He realizes that just like during his first encounter at the Green Chapel, he failed the test.

We may notice an interesting parallel between the poet's experiences in the Green Knight's castle and these he is encountering now. Previously, the Green Knight sent the lady (his wife) to test Gawain. Gawain failed the test. Now, the Prince has sent the maiden to test him. Again, he failed the test. However, there is quite a qualitative difference between these two encounters.

Afterwards, the poet finds himself back in his garden at the spot where he lost his pearl:

> I woke in that garden as before,
> My head upon that mound was laid
> Where once to earth my pearl had strayed.
> I stretched, and fell in great unease,
> And sighing to myself I prayed:
> "Now all be as that Prince may please."

At the end of his experiences, the poet comes to the following conclusion:

> To please that Prince had I always bent,
> Desired no more than was my share,
> And loyally been obedient.
> As the Pearl me prayed so debonair,
> I before God's face might have been sent,
> In his mysteries further maybe to fare.

The poet's state has changed quite substantially. His focus shifted from the maiden (the means) onto the Prince. And that was the purpose of his particular experience, i.e., to bring him into a direct link with a superior understanding and knowledge. He recognizes that the Prince is the Saki, his guide. Only with the Prince's help will the poet reach New Jerusalem.

The additional details about the structure of the invisible worlds are of particular interest to our discussion here. Namely, the invisible world is a multi-level structure. The structure described by the poet's experiences corresponds to that represented by the original pentangle. The poet introduced the pentangle in *Sir Gawain and the Green Knight*. For an ordinary mind, the invisible structure remains hidden; it is not perceptible. Additional faculties are required to perceive the overall design of the invisible worlds. And familiarity with this design will help us resolve the current problem of modern physics. But – we are not there yet.

A New Order of the Garter

O powerful love! that, in some respects,
makes a beast a man, in some other, a man a beast
(*William Shakespeare*)

The fate of the Order of the Garter followed the usual trend. It was only a matter of time before the newly introduced ideas degenerated and turned into an artificial and developmentally sterile set of rituals. After a little more than a century, the Order of the Garter's developmental function became fossilized. A new vehicle for the continuation of the process was needed. It was time to inject a renewed and more advanced impulse into Western society.

An interesting hint about the 16th century renewal of the Order of the Garter was inserted into one of Shakespeare's plays.

Shakespeare used his History Plays to illustrate the process that was initiated by the Order of the Garter. As indicated earlier, the original aim of the Order was to inject into society an element that ennobles man's mind. Symbolically, this element is referred to as "love." But as we have seen, this is not ordinary love. Instead, as phrased by one of Shakespeare's characters, this is a special love capable of making "a beast a man."

The first episode of the History Plays is set in the time of the reign of King John (1199-1216). The play *King John* illustrates the situation before the Order's formation. The following six episodes illustrate the process as it moves from the time of King John till the birth of Elizabeth I, i.e., the year 1533.

Shakespeare indicates that, as far as the development of "love" is concerned, there was no difference between the first and the following episodes of the History Plays. For whatever reason, "love" could not take root within the English royal channel. Unlike in his Italian plays, in the History Plays there is no "falling in love." The arrangements of the marriages are driven entirely by ordinary political and strategic motives. Somehow the element of "love" could not be activated; it remained latent. At one point, therefore, it was necessary to transplant this vital element into another milieu. According to Shakespeare's illustration, the transfer was implemented at the time of Queen Elizabeth I. The play *The Merry Wives of Windsor* symbolically illustrates this transfer. This play is set at the time of one of the Order's feasts which were held at Windsor Castle on the eve of St. George's Day. The play's final scene describes the Queen Fairy masque performed in Windsor Park. The masque may seem to be a rather silly and irrelevant folk ceremony. However, the masque is a symbolic illustration of quite an unusual event. It represents the new milieu for the process. An observant reader will notice that, like in the original "garter," thirteen named participants are taking part in the event. However, it is not the Knights of the Order but the ordinary folks from Windsor's middle and working classes who execute the event. Although the phrase "Our radiant queen" may be read as a reference to Queen Elizabeth I, it applies to the Fairy Queen, represented by Ann Page, the daughter of one of the Merry Wives. At the end of the masque, Ann and her lover are married. It is the first and only time in Shakespeare's plays set in England that "love" becomes the narrative's driving force.

The masque contains direct references to the Order of the Garter. For example, here is Mistress Quickly's instruction to the fairies:

> "Search Windsor Castle, elves, within and out:
> Strew good luck, ouphes, on every sacred room:
> That it may stand till the perpetual doom,
> In state as wholesome as in state 'tis fit,
> Worthy the owner, and the owner it.
> The several chairs of order look you scour
> With juice of balm and every precious flower:
> Each fair instalment, coat, and several crest,
> With loyal blazon, evermore be blest!"
> (*The Merry Wives of Windsor*, V.5)

"The several chairs of order" refer to the individual stalls assigned to the members of the Order in St. George's Chapel at Windsor Castle, the Knights' traditional meeting place. In the following quote, Mistress Quickly, who presides over the event, calls the fairies to form a ring and sing:

> "And nightly, meadow-fairies, look you sing,
> Like to the Garter's compass, in a ring:
> The expressure that it bears, green let it be,
> More fertile-fresh than all the field to see;
> And 'Honi soit qui mal y pense' write
> In emerald tufts, flowers purple, blue and white;
> Let sapphire, pearl and rich embroidery,
> Buckled below fair knighthood's bending knee:
> Fairies use flowers for their charactery."
> (*The Merry Wives of Windsor*, V.5)

The fairies are asked to use "emerald tufts" and flowers to write the salutation to the Saki: *Honi soit qui mal y pense*. The event is performed in a ring (a "garter") formed by the participants. The above quote refers to purple, blue, and white, i.e., the colors used by the Order of the Garter in the 17th century.

The ennobling impact of the activation of "love" is marked by the changes induced in the participants of the masque. Sir Hugh Evans is miraculously healed from his speaking idiosyncrasies; Mistress Quickly, who previously mixed the meaning of words, impeccably delivers her poetic message that also includes a phrase in French; Pistol, now happily in love with Mistress Quickly, delivers his lines flawlessly. Previously he could not correctly say a single sentence. These changes are similar to those associated with the presence of the *lady* of the Troubadours' songs. What is remarkable about this event is that it was conceived, directed, and implemented by Mistress Page and Mistress Ford, two housewives of Windsor. At that time, Shakespeare could not broadcast this sort of message openly. Therefore, he buried it within a seemingly absurd comedy in which Falstaff, the only knight among the play's characters, clearly demonstrates that he is incapable of falling in "love."

At this point, it may be interesting to notice another interesting detail related to Shakespeare's plays. Namely, in *The First Folio*, i.e., the first published collection of Shakespeare's plays, there is a page entitled "The Names of the Principal Actors" (see the attached image).

What is intriguing is that the actors' names are arranged in two groups of thirteen each – just like the composition of the original members of the Order of the Garter. This could be a subtle indication that the appearance of Shakespeare's plays had

to do with the renewal of the developmental process which was
initiated in England in the later part of the 16th century.

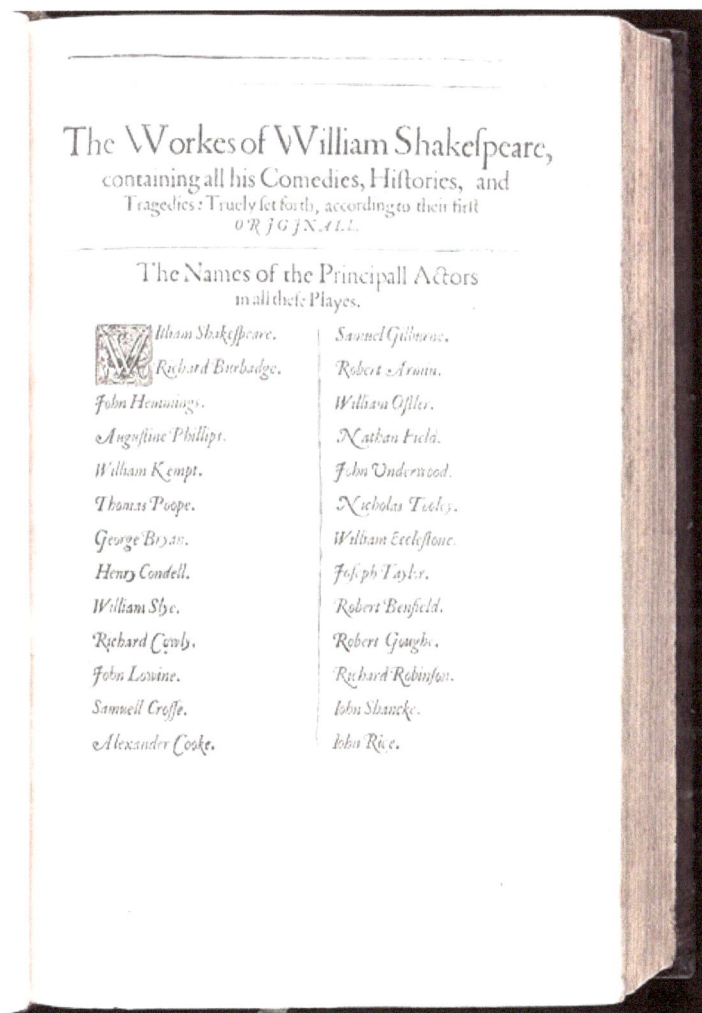

A page from Shakespeare's First Folio (1623)

***

As indicated earlier, the initiation of a new phase of the process is usually combined with the projection of a more advanced template of the human mind. As described in *Pearl*, more details of the multi-layer structure were disclosed when the Order of the Garter was formed. Therefore, we may expect that with the announcement of the formation of a new Order at the time of Elizabeth I, some further details about the structure would be released. Are there any hints about such details?

Yes, there are. They were also included in Shakespeare's plays.

Before Shakespeare's plays, the main characters in mystical poetry were a lover and his beloved, i.e., a couple. In other words, mystical poetry illustrated the various aspects of the interactions between a man and a *lady*. In Shakespeare's presentation, however, the impulse the lady represents – was greatly enhanced. The entire Shakespeare narrative illustrates the process that leads to a situation where such an advancement of the process was possible.

Shakespeare treated historical events as the manifestations of the state of the mind of a select group of people who were representative of a given geographical area at that time. Consequently, his various characters are a symbolic representation of the various faculties of the mind. Some faculties are ordinary, some are extraordinary, and others are at the point of gaining enhanced awareness. Such a composite state of mind determines what is possible and impossible; it defines its evolutionary potential and dictates the sequence of events. By using this allegory, Shakespeare could illustrate the process that led to the European Renaissance.[19] The first episode of Shakespeare's narrative takes plays in Troy (*Troilus and Cressida*), the commonly accepted cradle of Western civilization. Then it moves to pre-Roman Britain, ancient Rome, passes through the Middle Ages in France, England, Italy, and Central Europe, and concludes with the European Renaissance. So, the entire narrative is spread over a time of nearly three millennia. As the narrative moves from one place to another, the evolutionary impulse is gradually enhanced. In the beginning, it is represented by a lady. Then, it is transformed into two, three, and finally into four young women - who appear together. Their intended partners – represent the latent faculties. A marriage marks the successful activation of a new faculty. In the final episode of Shakespeare's narrative, there are four couples who are married at the same place and at the same time. They all together form a unity. Together they are all entangled.

This sort of group entanglement is of interest to us in the present discussion of science and consciousness. We will come back to it later.

---

[19] *Shakespeare's Elephant in Darkest England*, W. Jamroz, Troubadour Publications, Montreal, 2016.

Émilie, the Gambler

There are truths which are not for all men,
nor for all times.

(Voltaire)

Not so many people realize that the foundation of scientific determinism was laid down by a woman whose name has only recently begun to appear in the history of science. She was quite an extraordinary person who lived in 18th century France. She was better known as a mistress of Voltaire than as a talented physicist and mathematician. She was one of the first who tried to apply a perceptive approach to physics.

***

We are at the French court in Versailles in 1722. A crowd of nobles, courtiers, and foreign visitors are watching quite an unusual spectacle. It is a fencing duel between Jacques de Brun, the chief of the royal household guard and ... an attractive young woman. The woman is only 16 years old. She is tall, very fit, and very confident. She wields her sword like a hussar, squaring up furiously against her male opponent. The duel ended in a draw: the lady did not defeat her opponent, but de Brun was not able to defeat her either. They were both quite exhausted when they put down their swords.

It was in this fashion that Émilie du Châtelet gained her reputation among members of the court of Philippe II, the regent who ruled France after the death of Louis XIV. Émilie's parents sent her to live at the court, hoping that she would attract a wealthy suitor. But Émilie was not an ordinary woman. While at the court, she realized that she was surrounded by a ridiculous crowd of fops, gossipers, and gamblers. Émilie did not have time to waste on the peculiarities and hypocrisies of the French court. The duel was her way of letting everyone know she had other plans. Of course, for her mother, the duel was the worst possible outcome: instead of attracting a potential suitor, her daughter frightened them all away! Afterwards, her mother had wanted to send her to a convent. But no abbess would accept Émilie – obviously, Émilie's reputation had also reached those secluded domains. This meant that Émilie gained more time to do what she loved: study science.

However, in addition to being left alone, she also needed money to buy books. At that time, her father could not support her because his income had been reduced. This left Émilie with one option: gambling. By then, she had learned enough mathematics to use in gambling. Since she was so much quicker at calculating probabilities than anyone else, she could easily take advantage of her skills at the gambling tables. Gambling was quite popular among higher-class women. So, it was not unusual for her to join the ladies of the court at the card tables. Her partners at the gambling tables had no idea they were playing against one of the brightest mathematicians of the time. In a week, she was able to win more than two thousand Louis d'or. She would spend half of that money on new books. Yet, her father was not impressed. The prospect of getting a husband for her daughter was dimmer and dimmer. He knew that no great lord would marry a woman who was seen reading books every day.

At one point, however, Émilie realized that she would need a more stable income and a supportive environment to continue her studies. So, she decided to compromise her freedom by getting married. She married Florent-Claude, the marquis of Châtelet. Florent-Claude inherited enough money to guarantee his young wife a decent life.

Émilie learned Latin, English, Italian, and a good bit of Dutch and Greek. She translated Virgil and wrote commentaries on the Bible. However, Émilie's prime interest was science. At first, she studied Isaac Newton. Newton had created modern science, yet his work had been incomplete at his death. Émilie thought she knew what Newton's most important secret about the universe was.

In 1733 Émilie met Voltaire, a famous French writer, historian, and philosopher. At that time, she lived in Paris and was the mother of three children. Voltaire was 12 years her senior. Their intimate relationship began almost immediately. Their affair soon resulted in Parisians doing a lot of tongue-wagging, and Émile aided them. She was married, and instead of being discreet, she openly expressed her passion for Voltaire by kissing him full on the lips anywhere, anytime.

At that time, Voltaire was in trouble with the authorities for his writings, in which he criticized the class structure of French society, the Roman Catholic Church, and slavery. He was an advocate of freedom of speech, freedom of religion, and separation of church and state. Voltaire was expelled from France. His primary source of inspiration was the years he spent in England. It was there that he learned English. When he needed help with English pronunciation, he would make his way to the Drury Lane theater. The prompter would loan him a copy of that night's Shakespeare's play script, so he could practice the pronunciation while listening to the actors.

Émilie owned an estate called Cirey. As appearances were crucial, she thought Cirey might be the perfect escape from Parisian gossipers. Voltaire paid for the building's renovation. Émilie's husband - who was mostly away from home - happily welcomed this liaison. Sometimes he even stayed at the château with his wife and Voltaire. He and Voltaire became friends; they often would ride together and share dinners.

Émilie turned Cirey into a sort of private research center. It included a laboratory with scientific instruments. In addition to the laboratory, there was one of Europe's leading research libraries with nearly 21,000 books. Cirey was frequented by visitors who enjoyed conversations with their hosts."[20]

***

One of the visitors who came to visit Cirey was an English acquaintance of Voltaire. Voltaire had met him when attending Shakespeare's plays in London. Voltaire called him Yorick - after the character of a royal jester from Shakespeare's *Hamlet*. Yorick was quite an unusual man. Émilie was very attracted to him because of the things he knew and how he looked at life and the world. Also, he had quite an interesting approach to science. So Émilie spent many hours discussing with him a bigger picture of science.

Émilie talked about her interest in Newton and Leibniz during one of these discussions. She mentioned that Newton had shown that the universe was constructed in such a way that we would never know the underlying nature of any object. She believed that Newton had intentionally hidden the answers away

[20] The foregoing details about Émilie du Châtelet have been extracted from *Passionate Minds* by David Bodanis, Crown Publishers, New York, 2006.

deep within his convoluted text. On the other hand, that mysterious Baron Gottfried von Leibniz had suggested a detailed way God could control the universe. Furthermore, Leibniz seems to have proof of that. Upon hearing this, Yorick started to laugh, and he said:

"My dear Émilie, they both are wrong and right at the same time. One thing is to understand how science applies to the physical world. However, talking about God and how the world is controlled is entirely another matter."
Yorick's laughter annoyed Émilie.
"Why not? Everything in the universe is controlled by a set of energies. If we could only find out how these different energies are related and how to measure them – we would be in a position to know how the world is controlled."
"You are assuming," – Yorick continued, "that we can understand the mechanism of the world by simply applying the known laws of science to the invisible world. It does not work that way. It is the other way around. The laws of deterministic science reflect the laws that govern the invisible. Laws of physics are just a simplified projection, an approximation – of those invisible mechanisms. Therefore, it is impossible to infer the overall mechanism from the laws of physics. First, one must get familiar with those cosmic relationships – otherwise, you are in darkness; your target is beyond your grasp."
"You mean that even the greatest human intellects cannot grasp that?"
"Yes, this is what I am saying."
"You know – that I cannot accept that," – Émilie said with a disapproving smile.
"I will give you an example, an allegorical illustration."
"Great, go ahead!" – she heartily welcomed such a challenge.
"It is an example taken from that great English playwright – Shakespeare," said Yorick.

"Shakespeare?" - Émilie expressed her surprise.

"Yes, the very same with whom our dear friend Voltaire has such a … hate relationship."

This was true. Émilie knew that Voltaire hated Shakespeare. Yorick added: "from his play *Hamlet*."

"*Hamlet*?" – this was even more surprising to Émilie. Voltaire translated some of Shakespeare's plays, mainly some of the tragedies. Émilie knew that Voltaire was particularly dismayed with *Hamlet*. She remembered that in the commentary to his translation, Voltaire wrote that *Hamlet* was written by a drunken savage and that it was a vulgar and barbarous drama.

Yorick continued: "As you may remember, Voltaire wrote that in the graveyard episode in that play, the gravedigger talks nonsense, while Hamlet responds to his nasty vulgarities with silliness no less disgusting.

"O yes, I remember that" – Émilie still was in a state of surprise.

"You see, Voltaire's reaction to this very scene is a good illustration of what I am trying to point out."

"Really? And this has to do with my understanding of science?" – Émilie was obviously not convinced.

"Yes, it does. It may help you to get out of your current dilemma."

"All right, then make your proof!" – she added quickly.

"I shall," said Yorick – and he began his explanation:

"In this scene, the gravedigger and Hamlet represent two different states of the human mind; they both operate in different frameworks. Hamlet illustrates a clever, intelligent, but ordinary mind. The gravedigger, on the other hand, represents a much superior state of mind. This state allows him to see Hamlet's previous, current, and possible future experiences. He knows what Hamlet can and cannot do, what the most beneficial action would be, and …"

At this point, Émilie could not restrain herself from

interrupting with a sarcastic comment:

"You mean that nonsense-talking lunatic?"

"Let me finish – but your comment is quite a fitting element to what I am explaining," – Yorick riposted. Then, after taking a sip of wine from his glass, he continued:

"During that seemingly silly exchange, the gravedigger gives Hamlet several clues that should help him to recognize who the gravedigger really is. First, the gravedigger clearly indicates that he knows that he is talking to Hamlet, the Prince of Denmark. Hamlet, however, does not register that. In this scene, Hamlet appears to be clueless, even gullible. The rule is that – if there is no … minimum recognition of such hints, nothing can be done. And that was Hamlet's case. And the same applies to our approach to science."

"How so? I do not see any relevance between these two," – said Émilie.

It seems that Yorick was not surprised with Émilie's response, and he proceeded:

"Hamlet's state of mind corresponds to that of our rational, scientific mind. It can operate quite effectively within a certain framework, i.e., the physical world confined to time and space. However, his mind is oblivious to details that do not fit into this prefixed set of beliefs. In that sense – it is rather a dogmatic approach. The gravedigger's mind operates within a superior framework that contains a template of all things. So, he can see all and know how all things are linked and what is possible and impossible. That's a position you would have to arrive at – to be able to solve the problems that you are working on."

"You do not expect me to accept this?" – obviously, Émilie was not impressed. And she added:

"How can anyone extract such an understanding from the incoherent rubbish of that gravedigger?"

"You see, it is a matter of perception. The ordinary perception is limited to intellect. To accept something, one needs to rely on logic and rationality. However, the higher level of perception does not depend on such a clumsy thing as our intellect. At that level, one can perceive the overall picture. It is a much more comprehensive way of knowing things. So, let's go back to our gravedigger. The gravedigger gives a few hints to trigger Hamlet's curiosity. Namely, he mentions 'thirty years' and 'three and twenty years.' In the context of the episode at the graveyard, these numbers do not make any sense. They do not fit into that situation. These numbers seem to be mistakes or a sign of incoherence on the part of the playwright. This is why some commentators of Shakespeare accused him of lacking basic algebraic skills. Yet, Shakespeare was very precise with his numbers. But he did not use them to indicate how much time passed between the two events. In this scene, he used them as sort of markers or linking points. These numbers are like stitches that allow one to connect this particular episode with some previous ones. They indicate a link between the encounter at the graveyard and other events which took place earlier within Shakespeare's narrative. Linking these events together allows for recognizing an overriding design within which all of Shakespeare's episodes form a coherent whole. However, Hamlet would have to pay attention to what the gravedigger is saying. But Hamlet did not; he completely ignored the gravedigger's hints. In other words, he failed the test. As a result, he missed the opportunity to change the course of coming events. Therefore, the only task that remained for the gravedigger was to … dig the grave."

Yorick stopped for a moment, looked into Émilie's eyes, and added:

"By the way, Shakespeare was pretty good at numbers and mathematics. You may be interested to know that he was also quite fond of gambling."

This remark definitely triggered Émilie's interest.

"How do you know that?"

"From *Hamlet*. There is an interesting description of a gamble proposed by King Claudius."

"Really? I have not seen it in Voltaire's translation."

"You would have to read the original. Voltaire's translation is more about Voltaire's incomprehension than Shakespeare's design."

Émilie knew Voltaire wasn't much of a scientist; mathematics was too hard for him. Although Voltaire was bright, he had the attentiveness of a five-year-old boy. Instead of trying to understand, he preferred quick and entertaining quips that would contribute further to his fame of a great wit among salon-dwelling intellectuals.

Yorick carried on his explanation:

"Claudius organized a bet on the outcome of a swordfight between Hamlet and Laertes. The bet was Claudius' brilliant idea to avoid suspicion in case anyone discovered that Hamlet was killed by foul play. Claudius would look innocent when it was found that he had put a bet on Hamlet's winning in an earnest wager. So, he had to propose a wager which would make sense in accordance with the accepted gambling rules.

It was commonly known that Laertes was twice as good as Hamlet. Therefore, Claudius acting as the bookie, proposed a handicap wager to equalize the chances of Hamlet's winning – while maintaining statistical fairness. Now listen carefully - for this was a very well-conceived wager.

The swordfight was set for twelve passes. Because Laertes was twice as good as Hamlet, the handicap was that Laertes' eight hits were equivalent to Hamlet's four hits in a match of twelve passes. Such an outcome meant even money: no risk, no gain. Claudius, therefore, had to take a risk and bet higher on Hamlet to encourage other bettors. Otherwise, no one was going to participate in the gamble. Claudius bet that Hamlet would score at least five hits out of twelve passes. And he offered six Barbary horses as his stake. The other bettors matched his stake with six French rapiers. So, the stake had twelve items: six horses and six rapiers. By betting on Hamlet, Claudius was taking higher risk than the others. Therefore, he laid his bet as 'twelve for nine,' which meant that as his winning he could claim all the twelve items of the wager, i.e., the six rapiers, and he would keep his six horses. In the case of Hamlet's loss, the other bettors would get nine items of the wager, i.e., three horses, and they would keep their six rapiers. This was the meaning of the 'twelve for nine'."

"Whaa!" exclaimed Emile. "This is really good! How could Voltaire miss something like that?"

"You can ask him. I suspect that Shakespeare was quite familiar with Cardano's *Book on Games of Chance*. I saw that book in your library."

"Yes, indeed. I have learned quite a bit about gambling from that book. That Italian lad had some feeling of the sport!"

"Now, let me go back to the encounter at the graveyard. You see, Hamlet represents a bright intellectual; his understanding of the world around him corresponds to that of a scientist. On the other hand, the gravedigger can perceive the greater design within which all earthly events are connected. Therefore, he can help Hamlet, but Hamlet would have to demonstrate that he is capable of learning. This is why the gravedigger gives him hints. But, as I have said, Hamlet is too proud and too full of himself to even consider such a possibility."

After a short break, Yorick resumed his explanation:

"Right now, you are in a similar situation to Hamlet. The information you gathered would have to be interpreted within a greater design to find answers to your questions. Keep in mind that one part of the design is not complete without the other parts. Within that framework, answers and questions exist together as a one-to-one correspondence. You need to determine such a framework. Only then will you be able to find what you are looking for. You may think about it as a ladder that links together various energies, each more subtle than the next."

Émilie was silent. Yorick could see that she was deep in her thoughts.

Émilie knew, from her readings of biblical commentaries and other metaphysical sources, that there was a great tradition of hidden writing. In these writings, mystics, prophets, or others - who felt they had access to powerful knowledge - presented their findings in certain codes. "Is it possible," she asked herself, "that the same knowledge applied to science?" After a long period of silence, Émilie stood up and said" "Thank you, Yorick. I need to think about that. Have a good night and see you tomorrow."

Saying that she left the room.

***

Émilie conducted a series of experiments that led her to the determination of energy for objects in motion. She was the first person to formulate the equation for kinetic energy, where energy is equal to the mass of the object multiplied by its squared speed. The "squared speed," which appears in Einstein's famous equation $E=mc^2$, came directly from Émilie work. Afterwards, she came out with the law of conservation of energy. This law provided the foundation for modern scientific determinism. However, she extended the energy concept much further than modern science has considered. Namely, she postulated that all world movements, including the appearance and disappearance of cities, nations, and civilizations, must also be controlled by the law of energy conservation. Even though cultures were broken apart and their inhabitants dispersed, the total amount of "energy" would never change. For her, it was obvious that nothing ever entirely disappears, that nothing ever dies.

Later, she expanded her approach by trying to look at the physical world from a wider perspective. Namely, she considered the physical world as a derivative of a greater framework that was placed within the invisible. This led her to much more profound conclusions. She started to think about free will. She realized that another and more subtle form of energy must be involved in the operation of the human mind. She deduced that if we create new ideas, thoughts, and concepts, we inject a new form of energy that did not exist before. In other words, we have access to other energy sources beyond the grasp of deterministic science. But even these more subtle energies must also comply with the corresponding law of conservation. This means that the law of the conservation of energy, which she discovered, would have to be limited to a specific context, i.e., its application was limited to a particular framework. She started to apply this line of reasoning to the notion of matter.

Émilie presented her ideas about the notion of matter in her *Lessons in Physics* ("Institutions de Physique") published in 1740. Using her intuitive approach, she envisaged the matter structure that physicists would discover sometime later. She broke down physical matter into three parts: macroscopic objects, which are perceptible by sensory perception; atoms composing those macroscopic objects; and a sub-atomic substance, which is beyond the reach of sensory perception and measurements. (Today's physicists refer to this substance as a "quantum vacuum.") However, she cautiously emphasized that there was no way to know how many more levels of that "invisible" matter truly existed.

Émilie also predicted properties of matter that were to be discovered only 200 years later, i.e., within the framework of quantum mechanics. Namely, she implied that matter might appear in two forms: constant and variable. The first type was that of particles. The second, which she called "possible" objects, corresponded to the modern description of particles as waves. To exist, i.e., to appear as particles, the "possible" objects needed an external cause for their actualization. Hence, the existence of particles was the realization of the possibility inherent in "possible" objects. (In quantum mechanics, such realization corresponds to the "collapse" of the wave function).

Portrait of Emilie du Châtelet (by Marianne Loir)

Premature death did not allow Émilie to complete her work. She died on September 10th, 1749. Most of her work is still ignored or misunderstood. Today's commentators treat her writings as obscure rendering into metaphysics and disregard their scientific significance. Only her translation and commentary on Isaac Newton's *Philosophiæ Naturalis Principia Mathematica* got recognition among the scientific community. Published posthumously in 1759, the book is still considered the standard French translation.

***

Émilie's work indicates that the ideas initially disclosed within mystical poetry and other literary sources were slowly but surely reaching the scientific community. However, the task of advancing further such an approach was left to somebody else. There were still some pieces missing from the puzzle. Furthermore, more experimental data were needed to close the loop between deterministic science and the perceptive approach.

Waves, Resonance, and Vibration

If you want to find the secrets of the Universe,
think in terms of energy, frequency and vibration.

(*Nikola Tesla*)

So, how can the Troubadours' songs, the poems of the Pearl Poet, Shakespeare's plays, and some of Émilie du Châtelet's metaphysical renderings be of any use in resolving the problems of the big bang, elementary particles, dark matter, black holes, and quantum entanglement?

We will need one more thing before we can connect these two areas of inquiry, i.e., deterministic and perceptive. The missing link was provided by a seemingly non-related experiment that Ernst Chladni carried out at the beginning of the 19th century.

In the Chladni experiment, a disk-like plate is placed within an oscillating acoustic field. The plate is covered with grains of sand. When exposed to the acoustic field, spectacular shapes appear on the plate.

These various shapes are formed from grains of sand. The grains are aligned along those regions of the plate which do not vibrate. These non-vibrating regions correspond to the so-called nodes of the standing waves, which are induced in the plate by the acoustic field. The standing waves occur because the plate is in resonance with the acoustic field. An object, such as a plate, a container, a string, etc., has natural resonating frequencies, which are determined by the object's shape, geometry, the material of which it is made, and its internal structure. While in resonance with an external oscillating field, an object absorbs energy from that field and vibrates at its natural frequencies. A relatively small oscillation of the external field may induce large vibrations in the object. In other words, the resonance acts as a receiver and amplifier of the oscillations of the external field. The same principle is explored in the design of musical instruments. For example, the strings and body of a violin, the tube in a flute, or a drum membrane – are used as acoustic resonators.

Chladni patterns in a vibrating metallic plate[21]

It is important to realize that the patterns illustrated in the above figure are only a partial representation of the nodes. These patterns are only a tiny fraction of much more complex

---

[21] Chladni patterns published by John Tyndall in 1869
(https://www.physics.ucla.edu/demoweb/demomanual/acoustics/effects_of
_sound/chladni_plate.html).

structures. These more complex structures are three-dimensional.

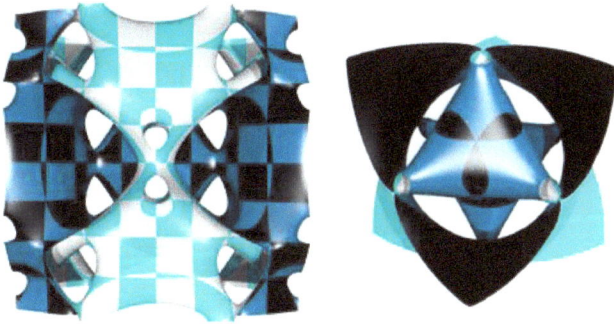

Three-dimensional Chladni patterns[22]

Three-dimensional nodes may be experimentally produced, or they can be mathematically modeled. Some examples of models of three-dimensional nodes are shown in the above illustration.

Chladni used these patterns to tabularize the entire spectrum of sounds. In other words, the Chladni patterns are equivalent to musical notes, i.e., a code used to represent sounds by graphical signs. For example, music notes for a Mozart's symphony are equivalent to hundreds and hundreds of such three-dimensional Chladni patterns.

---

[22] "Chladni Figures Revisited: A Peek Into The Third Dimension," M. Skrodzki, et al.
(https://www.researchgate.net/publication/319879322_Chladni_Figures_Revisited_A_Peek_Into_The_Third_Dimension).

Another form of Chladni patterns
(music notes for Mozart's Symphony No. 40)

Therefore, the Chladni patterns describe the structure of the acoustic field. The musical notes are another, more compressed form representing the acoustic field.

There is another form of manifestation of the acoustic field. This third form is even more compressed than musical notes. Great composers can perceive a piece of music in its entirety in a moment. Here is Mozart's account of such an experience:

> When I feel well and in a good humor, or when I am taking a drive or walking after a delicious meal, or in the night when I cannot sleep, thoughts crowd into my mind as easily as you could wish. Whence and how do they come? I do not know and I have nothing to do with it. ... Once I have my theme, another melody comes, linking itself with the first one, in accordance with the needs of the composition as a whole: the counterpoint, the part of

each instrument, and all these melodic fragments at last produce the complete work. ...It does not come to me successively, with various parts worked out in details, as they will later on, but it is in its entirety that my imagination lets me hear it.[23]

Mozart's experience is an example of a compressed manifestation of the acoustic field. This form is beyond the reach of ordinary senses such as hearing or sight.

Then, there is still another and even more subtle form of perceiving "musical harmony." Mystical poets often referred to this manifestation of music as "divine harmony." Here is an example of such writing:

From the Throne of the Eternal proceeds all Harmony.

The Music of the Spheres is the Divine Harmony,
and the Divine Harmony is the Law of the Manifestation
of the Unity,
for that Law being the only Law, must needs be the only
Harmony.

Behold the Light which illuminates the World;
behold the Harmony of Color of which it is composed.
It is the Manifestation to the Soul of Man
of the Glory of the Eternal, the Consciousness of his
Unity.

Hearken to the Voice of his Consciousness,
it is the Harmony of Music,
the sounding forth of the Glory which is manifested.

---

[23] *The Psychology of Invention in the Mathematical Field*, Jacques Hadamard, Princeton University Press, 1945, p. 16.

> To the Sage therefore Music and Color
> are but phases of the same expression,
> and to his Soul in ecstasy the joy of color
> and the sound of Music alight Divine emotions
> whereby his Soul feels for an instant
> as if the pale Reflection of the Eternal Consciousness
> were passing over it.[24]

The poem alludes to a level of manifestation of "harmony of music," which is much more complete than that described by Mozart. Shakespeare referred to it as "music that may not be heard." This level of perceptivity is equivalent to the experiences induced by the *lady* of the Troubadours and those described by the Pearl Poet. This means that even such a simple field as acoustic waves can be manifested on several levels. These levels form a kind of hierarchical structure.

At the top of this hierarchical structure is the ultimate source of harmony. The source is placed within an invisible field beyond the physical dimensions. In this form, the source corresponds to "music that may not be heard." At the first stage of its projection, the "music that may not be heard" is projected onto spacetime in the form of a miniaturized "big bang," i.e., a single point in space and time. It is this form that some exceptionally talented composers can perceive. At the next stage of its projection, the mini "bang" is translated into musical notes. This allows for the next stage of the projection, where musical notes are transformed into acoustic waves. Finally, the acoustic waves can be used to form various physical forms out of grains of sand.

[24] *The Mystic Rose from the Garden of the King*, Fairfax L. Cartwright, London, 1898, p. 275.

This kind of projection from the invisible world onto the physical world parallels the process of the formation of the universe.

The universe is in the shape of a disk. The shape of the disk is determined by spacetime. The disk-like universe is floating within invisible oscillating fields. This is very similar to Chladni's experiment. The difference is that in the case of the universe, instead of the various patterns formed from grains of sand, the invisible oscillating fields form all types of objects built from matter.

The invisible fields induce standing waves within spacetime. The standing waves bouncing back from the universe's boundaries form various nodes of physical objects. The shapes of these nodes are determined by the boundaries of spacetime and the "frequency" of the oscillations. When the frequency of the oscillations increases, more sophisticated shapes are formed. In the next stage of the process, the nodes are transformed into physical objects. It is in this manner that various physical objects appear in spacetime.

As we have seen in the discussion of *Sir Gawain and the Green Knight*, the structure of the human mind may be represented by a pentangle. In its original form, the stars of the pentangle represented the various layers of the human mind. In accordance with the formula "As above, so below," the pentangle can also be used to illustrate the structure of the physical universe. The enclosed diagram shows the pentangle of matter, i.e., it outlines the five main regions of the structure of matter.

112

The pentangle of matter
- External star: elementary particles
- First inner star: atoms, molecules
- Second inner star: galaxies
- Third inner star: planets
- Inner dot: the Earth

The outside star of the pentangle represents the elementary particles. The next region contains atoms and molecules. These are followed by galaxies and then by planets. The dot in the center of the pentangle corresponds to the planet Earth.

The entire pentangle of matter is enclosed within invisible oscillating fields. These fields belong to the world of templates, i.e., they are outside of the physical dimensions. The templates are the source of the various nodes that appear within spacetime.

This diagram may also help grasp better the concept of the big bang. Namely, at the beginning of time, the source of matter was in the form of an infinitely tiny dot (just like the source of "music that may not be heard"). This dot contained the compressed forms of the templates of all physical objects. The big bang refers to the event when that tiny dot rapidly expanded. Then, all the templates were projected onto spacetime and appeared in the various forms of the matter. This means that all templates of physical objects existed before the big bang. Marie Howe intuitively perceived the existence of such templates when she wrote in her poem:

> …
> but only a tiny tiny tiny tiny dot brimming with
> *is is is is is*
> All   everything   home

Similarly, visual artists have also been engaged in the discussion of some of the topics of modern physics. For example, the attached photo shows a sculpture by Gianfranco Meggiato, a contemporary Italian sculptor.

"Quantic Sphere" by Gianfranco Meggiato
on display in Sicily, the Valley of Temples
(photo credit: Dominique Hugon, 2021)

The sculpture, called Quantic Sphere, was temporarily installed in Sicily's legendary Valley of Temples. The sculpture is a representation of the universe. In it, the universe is represented by a sphere that is floating within a web of rather sophisticated fields. The description attached to the sculpture reads:

> Quantum physics or mechanics, also known as quantum theory, represents the evolution of traditional physics and focuses on the behavior of matter in the microcosm. Composed of billions of particles, matter contains a defined quantity of 'quanta' connected to each other. This brings about the consideration of a perfectly organized universe in which everything is correlated and in perfect harmony in space and time and everything descends from One, from the Principle.

In his interview with *Forbes*, the artist further explained his vision of the world:

> I was keen to approach the Valley of Temples in a new way, taking a journey into the future by connecting these sites to quantum physics while also seeking to merge art and science as Leonardo Da Vinci did.[25]

As artists are more open to intuitive understanding, it is easier for them to perceive an overall picture of the universe. As we can see, the artistic perception of the physical world is very close to the actual structure of the universe.

---

[25] From an interview in *Forbes*, Aug. 16, 2021
(https://www.forbes.com/sites/stephanrabimov/2021/08/16/there-is-no-future-without-memory-sculptor-gianfranco-meggiato-on-the-social-legacy-of-art/?sh=44361f755ce4).

Entanglement

All matter originates and exists only by virtue of a force
which brings the particle of an atom to vibration and
holds this most minute solar system of the atom together.
We must assume behind this force the existence of a
conscious and intelligent mind. This mind is the matrix of
all matter.

(Max Planck)

The invisible fields within which the universe is floating –
are not homogeneous. Instead, they form multiple zones which
contain the templates of the various physical objects. These
various zones operate at different "frequencies." Each of these
frequencies carries the templates of objects, which are then
projected onto spacetime and appear in the physical world.

Elementary particles are the smallest objects in the physical
world. They appear in the area which is very close to the
boundary with the invisible fields. Because of their closeness to
the invisible fields, elementary particles are unstable. They may
appear both as particles and as waves. In physics, they are
classified as quantum particles. The laws of quantum mechanics
govern them. They serve as the building blocks for more
complex systems.

Let's briefly review the current state of physics of elementary
particles because it is there that we will find further hints about
the overall structure of matter.

The physics of elementary particles aims to develop a so-called standard model. This model describes all particles that physicists have seen in the experiments carried out in particle colliders. Since the 1930s, scientists have been using particle colliders to gain insights into the structure of matter. These machines are the most powerful experimental tools that are available to science. For example, the Large Hadron Collider at the European nuclear research center (CERN), near Geneva, consists of a 27-kilometer ring of superconducting magnets. Colliders accelerate particles nearly to the speed of light and then smash them. This allows physicists to study what comes out of the collisions. In this manner, the entire family of the smallest particles of matter was discovered.

There are two types of elementary particles. One group is called "matter particles." These are particles from which matter is built. Among matter particles are electrons, baryons (protons and neutrons), and quarks. However, quarks are not really particles. Instead, they are fractions or fragments of particles. Quarks always appear as an assembly that forms other particles. They are never found in isolation. They are the constituents of other particles. For example, three quarks are needed to make up a proton or a neutron. Quarks are the smallest known elements of matter. It was demonstrated that quarks are smaller than 10^{-18} (i.e., 0.000000000000000001) meter. Colliders can measure the size of particles down to 10^{-20} meter. Therefore, the size of quarks is nearly at the limit of the colliders' resolution.

At this point, it is interesting to recall that according to the uncertainty principle, there is a size limit to a particle of matter. This limit is known as the Planck length. It has been determined that the Planck length is 1.6×10^{-35} meter. No element of matter can exist below that limit. So, there is still a lot of unexplored space between the size of a quark and the Planck length.

Therefore, there is still a possibility of discovering a series of even smaller elements of matter.[26]

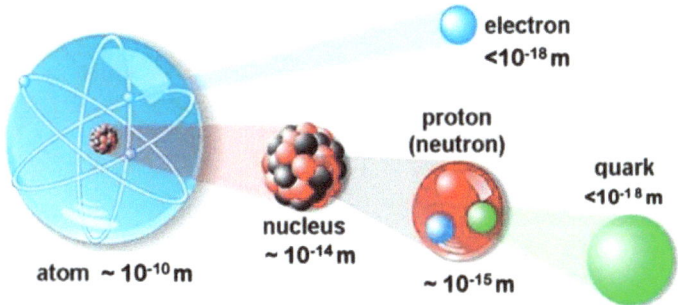

electron
$<10^{-18}$ m

proton
(neutron)

quark
$<10^{-18}$ m

nucleus
$\sim 10^{-14}$ m

atom $\sim 10^{-10}$ m

$\sim 10^{-15}$ m

The smallest constituents of matter[27]

The second group is the so-called "force particles." These particles mediate between forces of nature and matter particles. Physicists identified three forces that are prominent in the sub-atomic region, i.e., electromagnetic, radioactive, and nuclear. Therefore, there are three types of "force" particles: photons, gluons, and bosons. The photons convey the electromagnetic force; the co-called W and Z bosons mediate the radioactive force; the gluons carry the nuclear force.

For example, a proton is composed of three quarks which are "glued" together by gluons.

---

[26] In July 2022, scientists at the European nuclear research center (CERN), reported the discovery of never-before-seen combinations of four and five quarks.
[27] https://slideplayer.com/slide/12700821/

To complete the model, physicists needed one more particle, which they called the Higgs boson or the "God particle." This particle was needed to give mass to the matter particles. The discovery of the Higgs boson was announced in July 2012. This discovery fulfilled a 45-year-old prediction, captured worldwide attention, and thrust physicists into euphoria.

However, even with the discovery of the Higgs boson, the standard model does not include gravity, i.e., the fourth force of nature. Gravity is not compatible with quantum mechanics. It is believed that there must be a particle responsible for gravitation. This particle was named the "graviton." But this particle has not yet been found. Without the graviton, the problem of compatibility of gravity with quantum mechanics cannot be solved. And this is a serious problem for elementary particle physicists. Despite working intensely on the standard model for nearly a century, physicists have reached a dead end. The standard model cannot be the ultimate theory of matter because it leaves out the force of gravity.

***

Atoms, molecules, and minerals appear in a region farther away from the boundary of the invisible fields. They are built primarily from protons and neutrons. Protons and neutrons form what is called "baryonic matter." Because baryonic structures are located further away from the boundary with the invisible fields, they have different properties than quantum particles. They do not manifest duality anymore; they are stable.

Baryonic structures are subject to the electromagnetic force and gravity. In larger baryonic structures, gravity overrides the laws of quantum mechanics. And this is a crucial observation which, somehow, was ignored by physicists. Driven by their deterministic *belief*, physicists assumed that all physical objects must comply with the laws of quantum mechanics. Yet, there is a fundamental difference between quantum particles and baryonic objects. Quantum particles are basically waves. They behave and travel as waves. A wave cannot be confined to a single point; its state cannot be precisely determined. It can only be approximated. So, it seems somehow obvious that the wave function describing particles is probabilistic and uncertain. Baryonic matter, on the other hand, corresponds to the state of permanently or quasi-permanently collapsed wave functions. In this "collapsed" state, baryonic matter has already been "measured" or "observed," i.e., it has been converted from its wavy form. Therefore, the previously mentioned "quantum measurement problem" does not apply to baryonic matter. In other words, the conclusions formulated by quantum mechanics do not apply to cats and galaxies. The quantum ambiguity appears because physicists insist that all objects are waves and behave in the same way as quantum particles. It was this deterministic insistence that became the source of all those paradoxes with cats, gunpowder, and other supposed "mysteries" of quantum mechanics.

As we will see, a different mechanism governs baryonic matter.

***

The smallest structures in the baryonic region are atoms. An atom consists of a cloud of electrons that orbit around a nucleus. A nucleus is built from protons and neutrons, i.e., baryons. The overall structure of an atom is held together by the electromagnetic force. What is of particular interest to our discussion is that each electron acquires unique characteristics within an atom. More specifically, four different quantum numbers describe each electron. These quantum numbers are called: principal, azimuthal, magnetic, and spin (see the enclosed illustration).

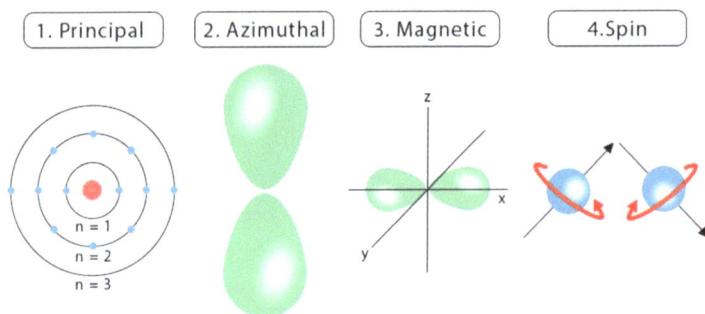

Quantum numbers[28]

Principal: determines the distance of the electron from the nucleus
Azimuthal: determines the shape of the orbit of the electron
Magnetic: determines the orientation of the orbit of the electron
Spin: determines the orientation of the axis of rotation of the electron

In accordance with the Pauli exclusion principle, two electrons within an atom cannot have identical values of those four quantum numbers.

What this means is that within the structure of an atom, each electron is unique. It can be said that within an atom, all electrons are "aware" of their own state and the states of all other electrons. This is what is called quantum entanglement. In other words, the Pauli exclusion principle is the consequence of quantum entanglement.

---

[28] https://www.chemistrylearner.com/quantum-numbers.html

In the scientific literature, it is often stated that it is possible for a pair of electrons to preserve their entanglement, even if they are light-years apart. Measuring a property of one of them, for example its spin, instantaneously affects the measurement of its mate's spin. An electron can have only two values of its spin, i.e., spin up or spin down. If one entangled electron's spin is up, then another electron would have its spin down. This may be compared to the flipping of two coins. When flipping two coins, there are three possible options, i.e., two heads, two tails, or one tail and one head. In the case of two entangled electrons, however, if one electron's spin comes up, the other would invariably be found with its spin down. This odd connection between two electrons seemingly breaks the universe's known laws. Yet, laboratory experiments confirm its reality every day. Photons also display similar behavior. In 2022, the Nobel Prize in Physics was awarded to a trio of physicists whose experiments over the years had demonstrated the existence of this "spooky action at a distance." However, the Nobel Prize committee was cautious with the way they phrased their nomination. The nomination emphasized that the award was not so much for explaining the nature of entanglement – but rather "for experiments with entangled photons." Today, this phenomenon is as mystifying as it was 75 years ago.

The description given in the previous chapter indicates that it will not be possible to explain the nature of quantum entanglement as long as the invisible fields are not included in the model of matter.

Quantum entanglement of particles is facilitated through their common template. This template is placed within those invisible fields, i.e., outside of spacetime. Therefore, the connection is not so much between particles. There is no physical link between entangled particles. Instead, the electrons are connected indirectly through their common template. This indirect connection with their common template facilitates that "link without a link." It is this indirect or "non-local" connection that is manifested by quantum entanglement.

It is through their common template that particles form a whole, i.e., one entity. Quantum entanglement is a manifestation of that "wholeness," i.e., a link between visible and invisible. This is why observing an instantaneous connection between two particles is possible.

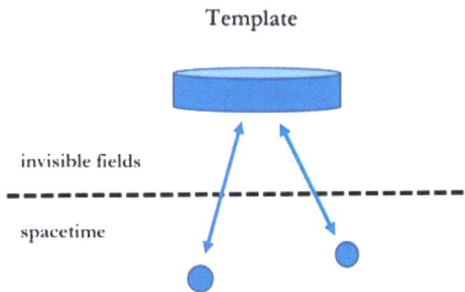

Quantum entanglement of two particles (the link without a link)

The existence of a common template also helps to explain other forms of entanglement. Namely, it was demonstrated that quantum entanglement could be extended from a pair of entangled particles to other particles. For example, let's assume that two entangled particles have been separated and placed at different locations. It is possible to entangle a third particle with either of the two. It does not matter which of the two is entangled to the third particle – the net result is that all three particles become entangled. This means that the third particle becomes entangled with a particle with which it has never been in contact. They are entangled even though they never met! It is a sort of group effect: an individual member of an entangled assembly can entangle another particle. As a result, the newly entangled particle becomes entangled with all members of the entangled group. They are all linked together through their common template placed outside of spacetime. It is the nature and mechanism of this mysterious effect that causes so many troubles for physicists. However, as soon as the concept of the common template is grasped, then all the mystery of entanglement disappears.

At this point, it is interesting to point out that the above-mentioned "group effect" was described for the first time in its symbolic form in Shakespeare's plays. As mentioned earlier, before Shakespeare's plays, the main characters in mystical poetry were a lover and his beloved, i.e., a couple. In Shakespeare's presentation, however, the lady was … split into four aspects. All these four aspects are entangled – just like a group of several particles can be entangled together. We should add that the possibility of group entanglement of photons makes it possible to construct quantum computers. So, it would not be too far-fetched to say that the overall concept on which quantum computers are based - was first indicated in Shakespeare's plays.

Conscious Crystals

What appears to be truth
is a worldly distortion of objective truth.

(Hakim Sanai)

From elementary particles and atoms, we now move to the region of matter that corresponds to molecules, minerals, and crystals. These systems, particularly crystals, display some amazing properties, which can also be explained with the help of the previously discussed model of the structure of matter.

A crystal is a solid material whose components are arranged in a highly symmetric and orderly repeating pattern. The components of crystalline lattices are either atoms, ions, or molecules. Two types of symmetries are observed, translational and rotational. Translational symmetry means that the entire structure repeats itself when a cell is shifted up, down, left, or right. In rotational symmetry, a cell looks the same after a rotation. The interesting thing is that, for many years, it was believed that only certain rotational symmetries could be observed in nature. Namely, despite thousands of possible atomic arrangements, crystals could have only two-, three-, four- or six-fold rotational symmetry. This means that the shape of crystal cells would be the same only when rotated by 180-degree, 120-degree, 90-degree, or 60-degree. It seemed that, for some unknown reasons, nature did not allow the formation of any other arrangements. And the known theory of crystal growth could not explain the crystal lattices' preferences for these symmetries.

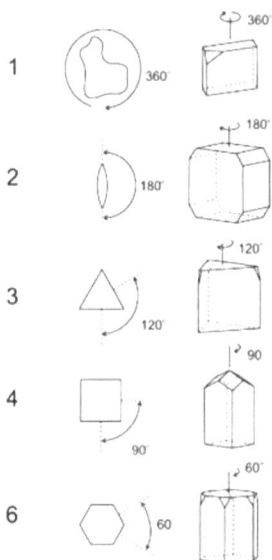

Rotational symmetry[29]

The accepted theory of crystal growth assumes that the lattice structure is formed by molecules searching for a configuration that requires the least energy to assemble itself. This is schematically illustrated in the following drawing:

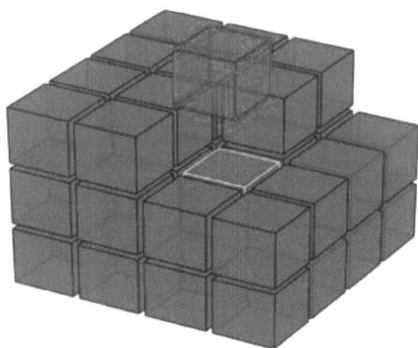

Crystal growth[30]

[29] https://opengeology.org/Mineralogy/10-crystal-morphology-and-symmetry/
[30] https://en.wikipedia.org/wiki/Crystal_growth

The drawing shows a simple structure formed from cubic molecules. As we can see, the top layer of the lattice is incomplete, as only ten of the sixteen positions are occupied by particles. A new molecule (shown with red edges) joins the crystal. It is joining the lattice at the point which requires minimum energy. Such a point is in the corner of the incomplete top layer (on top of the molecules shown with yellow edges). In this position, the energy necessary to hold it in place will be minimum because the molecule will be supported by three neighbors (one below, one to its left, and one to its right). All other positions have only one or two neighbors. According to this theory, crystal growth is realized by adding molecules, one at a time.

However, this understanding of crystal growth was challenged when a new type of crystalline structure was discovered. These new structures became known as quasicrystals because when they were first discovered, these materials were thermodynamically unstable: when heated, they would transform themselves into regular crystals. Later, many stable quasicrystals were discovered, making it possible to produce large samples for their study and applications.

Quasicrystals are types of crystal arrays of molecules with a highly ordered structure – but this structure is not periodic. A quasicrystalline pattern can continuously fill all available space but lacks translational symmetry. This means that, unlike regular crystals, a shifted copy will never match exactly its original.

The analogy of tiling a bathroom floor is often used to explain the properties of these materials. If one wants to cover a bathroom floor, then only tiles of specific shapes fit together. Namely, only tiles shaped as rectangles, triangles, squares, or hexagons can be used for this purpose. Any other simple shape of tiles would leave a gap. This feature of tiling is consistent with the natural symmetry of regular crystals. However, it was discovered that quasicrystals contain an ordered structure, but the patterns are more sophisticated and do not recur at regular intervals. Instead, quasicrystals appear to be formed from several different structures assembled in a nonrepeating array. As a result, it is possible to have a five-fold or ten-fold rotational symmetry. It turns out that any rotational symmetry becomes attainable. Therefore, with quasi-periodicity, a whole new class of solids is possible.

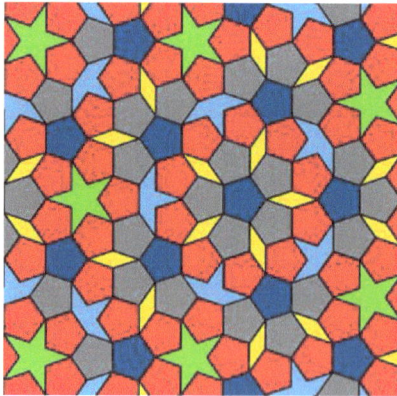

Penrose tiling (a set of six prototiles: green, red, yellow, light blue, grey, and dark blue)[31]

The interesting thing is that the possibility of the existence of such quasi-periodic structures was first theoretically studied by mathematicians. Some theoretically developed structures were described in a paper published by Roger Penrose in 1974 (see the above illustration). These unusual structures became known as Penrose tiling. Interest in these structures was further stimulated in 1982. That year, Danny Shechtman of Israel's Technion University discovered that certain metallic alloys displayed a very unusual symmetry. While examining a mix of

[31] https://en.wikipedia.org/wiki/Penrose_tiling#/media/File:Penrose

aluminum and manganese, an alloy with potential uses in aerospace, Shechtman found that the atoms were arranged in a pattern that could not be repeated by any translation. At the same time, the atoms in the sample seemed to be arranged in a pattern that had a five-fold rotational symmetry. The known laws of chemistry forbade such a structure. Shechtman named these structures quasicrystals.

At first, Shechtman was afraid to publish his data because of the hostility of other scientists. When he reported his data, he was a subject of ridicule and was repeatedly lectured by others about the basis of crystallography. For example, Linus Pauling, the two-time Nobel Laureate, is on the record as saying: "There is no such thing as quasicrystals, only quasi-scientists." It took some years before other researchers confirmed Shechtman's findings. In 2011, Shechtman was awarded the Nobel Prize in Chemistry. The Nobel Committee stated that "His discovery of quasicrystals revealed a new principle for packing of atoms and molecules" and pointed out that "this led to a paradigm shift within chemistry."

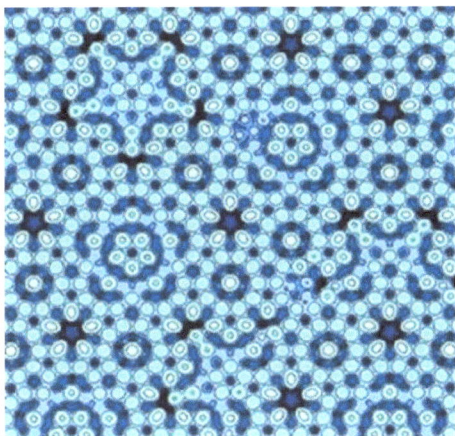

A quasicrystal of silver aluminum (a five-fold symmetry).[32]

---

[32] https://upload.wikimedia.org/wikipedia/commons/5/5d/Quasicrystal1.jpg

Although the pentagons in the above illustration can't fit together as squares and triangles do, other atomic shapes fill the gaps.

In 2013, the Post of Israel issued a stamp dedicated to quasicrystals and the 2011 Nobel Prize.

A stamp honoring the discovery of quasicrystals

There is another interesting aspect of these quasi-periodic structures. Namely, the same types of patterns were found in the design of the medieval Islamic mosaics of the 13th century Alhambra palace in Spain and the 15th century Darb-i Imam shrine in Iran. In 2007, the mosaic design patterns were deciphered by Peter Lu of Harvard University and Paul Steinhardt of Princeton University. The scientific community was quite surprised to find out that these designs included Penrose tiling. This mosaics' design has helped scientists understand what quasicrystals look like at the atomic level. In those mosaics, as in quasicrystals, the never-repeating pattern arises from the irrational number that is at the heart of its design. This number is the famous number known as the golden ratio.

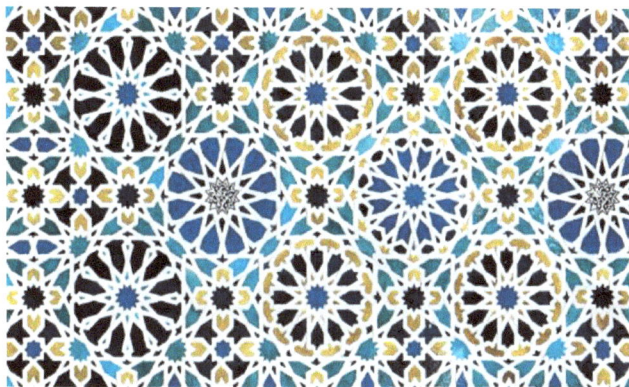

Tilework in the Alhambra palace, Spain

Archway from the Darb-i Imam shrine with quasicrystal-like patterns

The first human-made quasicrystals were by-products of the Trinity nuclear bomb test in Alamogordo, NM, in 1945. The test produced icosahedral-shaped quasicrystals. However, they were unnoticed at that time. They were identified only in 2021. They are the oldest known human-made quasicrystals. The same quasicrystals were also found in a specimen known as Khatyrka, a meteorite found in Russia.

The investigation of quasicrystals showed that their assembly could not be achieved by simply adding molecules one at a time.

It was suggested that the process of growing quasicrystals was necessarily non-local. Penrose proposed a possible mechanism for their formation, during which an unidentified non-local quantum-mechanical element is involved.[33] Such a process requires the cooperative effort of a large number of molecules all at once. Although a bit vague, this proposal clearly points out that the growth of quasicrystals cannot be achieved by adding molecules locally one at a time. This means that the existing theory of crystal growth cannot satisfactorily explain the process.

However, the main conclusion of the study of quasicrystals is a feature that scientists carefully avoid spelling out. Namely, the individual molecules forming quasicrystals are clearly "aware" of the overall design of the entire lattice. In other words, the molecules are entangled. Secondly, and equally important, this type of "awareness" differs from that observed in the case of quantum entanglement. In the case of quantum entanglement, there is a kind of awareness between electrons (quantum particles) and an atomic structure. In the case of quasicrystals, the molecules (baryonic particles) are aware of the entire crystal structure. The crystal structure is an example of a solid. Therefore, in this case, it is entanglement between baryonic matter and the corresponding template placed within a higher zone of the invisible fields. The baryonic template is this unidentified non-local element; but this is not a quantum mechanical element.

Quasicrystals demonstrate entanglement on the scale of solids. In other words, quasicrystals are examples of "conscious" solids. And this statement is an appropriate answer to the question asked by Marie Howe in her poem quoted at the beginning of this book: "Can molecules remember it?"

---

[33] *The Emperor's Mind*, Roger Penrose, Oxford University Press, New York, 1989, p. 436.

Yes, molecules can remember a few things. They are also "aware" of their rather sophisticated structure.

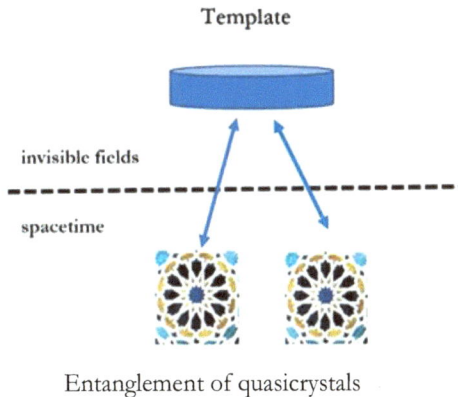

Entanglement of quasicrystals

Our discussion of the structure of crystals has taken us along the spectrum of matter from elementary particles to systems built from an assembly of molecules. This means that we are within a region that is controlled by a different zone of the invisible fields. In this zone, the templates of the baryonic structures are placed.

The force involved into the projection of baryonic structures onto spacetime is different than in the case of quantum particles. Therefore, there is another form of "awareness," which may be called baryonic entanglement. Baryonic entanglement is manifested by the "awareness" of molecules of the overall design of the structure they are part of. And this kind of relationship may be observed among all baryonic objects.

Complex Matter

Now over this water you wish to fare:
By another course you must that attain.

(The Pearl Poet)

There is another aspect that needs to be explained before this model of matter is completed. It relates to the interaction or communication between the various templates and their physical derivatives. Because there must be a mechanism or a force to project the invisible templates onto the physical spacetime. Such a force has not yet been identified by physicists. Some particles must facilitate such a projection, just like "force particles" mediate between physical forces and other elementary particles. However, these cannot be ordinary particles. These particles must be of an entirely different kind.

At this point, the *lady* of the Troubadours and the Pearl Poet's poem may greatly help.

Let's recall that the lady of the Troubadours provided a bridge allowing to … cross from the ordinary physical world into the invisible one. This is why the lady's physical form was not described by the Troubadours. She was "imaginary." Instead, those love songs described her effect on the lover. The heroine of *Pearl* went much further in providing more details about the structure of the invisible world. Namely, she indicated that there was a certain boundary between these two worlds. The boundary was in the form of a stream (see the illustration on page 71). Moreover, she could cross the boundary and appear in both worlds.

The maiden allowed the poet to see the boundary and the realm beyond it. However, the poet was not able to cross the stream:

> Now over this water you wish to fare:
> By another course you must that attain.

The boundary marks a transition between the visible and the invisible. It separates the phenomenal world from the hidden world of templates. In *Pearl*, New Jerusalem is on the other side of the water. New Jerusalem represents the world of templates.

Today's physicists are in an analogous situation to the Pearl Poet. Crossing "this water" is equivalent to crossing the Planck limit and entering a zone within the world of templates. The only difference between these two examples is that they are placed in two different zones of the invisible world. Namely, all that is next to the Planck limit belongs to the lowest zone, while New Jerusalem belongs to the highest zone of the world of templates. This means that the templates of quantum particles are in the lowest zone of the invisible world.

When physicists attempted to simulate "nothingness," they arrived at the crossing of "this water." First, let's recall that "nothingness" or a quantum vacuum is a region with some invisible bubbling fields within which various shapes of future elementary particles pop in and out. Physicists refer to them as virtual particles. The following image is a screenshot of a video showing such a simulation of virtual particles.

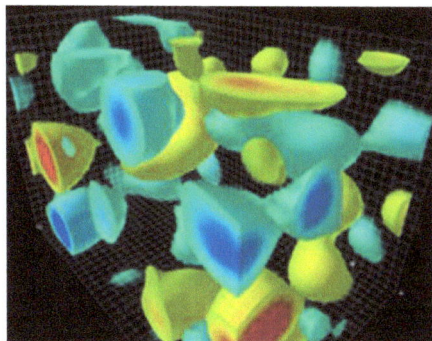

A visualization of quantum nothingness[34]

Modeling "nothingness" is quite a remarkable moment in the history of science. Although not recognized by physicists as such yet, it is the first attempt at modeling something that is … imaginary, i.e., something that does not contain physical matter. It is a first step leading to perceptive science.

Physicists still believe that these bubbling fields of "nothingness" are a physical entity, and together with elementary particles they constitute a complete system needed to explain the creation of matter. Such an understanding corresponds to the Pearl Poet's attempt to cross the water. The maiden's response to the poet applies equally to elementary particle physicists: "by another course, you must that attain." In other words, there is another way to bridge elementary particles with the source of matter.

The comparison with the experiences described in *Pearl* allows us to extrapolate more information about the new family of particles needed to complete the model of matter. As mentioned earlier, these particles are needed to link the physical world and the world of templates. In other words, such particles must link both worlds.

---

[34] https://en.wikipedia.org/wiki/Quantum_fluctuation

Let's go back to the experiences of the Pearl Poet. The overall situation illustrated in *Pearl* is an encounter between physical and non-physical worlds. The main point of this description is that the heroine of *Pearl* serves as a link between these two worlds. The important thing to notice is that her appearances are different, depending in which world she is present. When in the physical world, the poet sees her in the shape of a precious *pearl*. When in the other world, she appears as a *maiden*. And this hint helps to determine the form of the missing particles that link the invisible and the physical worlds. The hint is that such particles must exist in two forms at the same time. One form is physical; the other is "imaginary," i.e., it is placed outside of the physical dimensions.

How could this be possible?

It turns out that such a scientific device which consists of two forms, one "real" and the other "imaginary," has been known since the 16th century. Cardano, the Italian mathematician from whose book Émilie du Châtelet learned her gambling skills, discovered it. This device is known as "complex numbers." A complex number consists of two parts. One part is called "real," and the other is called "imaginary." In the notation of the complex numbers, the complex entity is the sum of these two parts:

$$\text{Complex entity} = \text{Real part} + \text{Imaginary part}$$

Complex numbers helped to resolve many calculations in mathematics and physics. However, so far, complex numbers have never been used in an application where one part was truly imaginary, i.e., belonged to the invisible world. It seems that up to now, the complex numbers were like a solution waiting for a problem. A similar situation occurred when the laser was invented. After its invention, the laser was also called "a solution waiting for a problem" because there were no apparent applications that would need such a device. It took some time before laser applications were developed and mastered. The same situation is now with complex numbers. They have been applied widely, but they were not really needed. It is only now that a problem has appeared that will require them.

The particles which communicate between the physical world and the world of templates are "complex." They consist of two parts. A complex particle must have a real component and an imaginary one. This is why there is so much difficulty explaining the origin of elementary particles and matter. Therefore, there should be no surprise that, as stated by Professor David Tong, "we are still a long way from understanding." Because, in this case, the "imaginary" component is genuinely imaginary, i.e., it does not contain physical matter. This means that it cannot be measured directly. On the other hand, the "real" part can be measured. But the complex particle cannot be complete without its imaginary mate.

Now we may realize that what physicists call "nothingness" or quantum vacuum is an interface between the visible and the invisible. "Nothingness" contains the imaginary parts of the complex particles. The imaginary parts serve as an interface between the world of templates and the physical matter. In other words, those "bubbling fields" contain the imaginary parts of complex particles which allow to complete the link between the visible and the invisible.

Now, we have the following question: What constitutes the "real" part of these complex particles?

An imaginary part must be coupled to a corresponding "real" part to form a complex particle. Therefore, the "real" part

140

must be like a … fragment or a fraction of particle. Are there any such types of particles among those discovered by physicists within the quantum world?

Yes, there are.

As indicated earlier, quarks are examples of fragmented or fractional particles that have been known for some time. There are six types or flavors of quarks, i.e., up, down, top, bottom, charm, and strange.

The existence of quarks was suggested by Murray Gell-Mann, an American physicist who received the 1969 Nobel Prize for his work on the theory of elementary particles. The fanciful name "quark" was taken by Gell-Mann from a line in James Joyce's novel *Finnegan's Wake*, "Three quarks for Muster Mark."

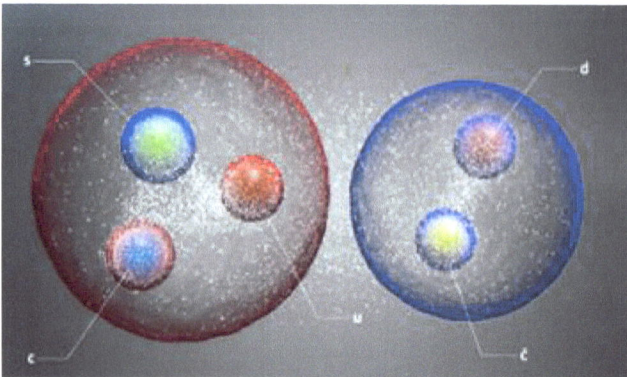

A newly discovered pentaquark[35]

So far, quarks have been treated by physicists like any other particles.

---

[35] https://www.iflscience.com/three-new-particles-discovered-by-the-large-hadron-collider-64335

Quarks possess some very unusual properties. For example, their electric charge is also fractional, i.e., one-third or two-thirds of that the electron. Their fractionality is nicely illustrated on the above image of a pentaquark, a particle discovered at the Large Hadron Collider in July 2022. The pentaquark consists of five quarks, i.e., a charm quark (c), a charm antiquark ($\bar{c}$), an up quark (u), a down quark (d), and a strange quark (s).

Their fractional nature indicates that quarks, or similar particles still to be discovered, are the partners of those imaginary parts that reside in "nothingness." Therefore, a complex particle consists of a quark-like particle and an imaginary part. And it is these complex particles that facilitate the formation of physical matter. It is through *quarks* that all other particles are linked to the invisible templates.

This means that only quark-like particles are linked directly to the quantum vacuum. Although fractional, they are the first forms of physical matter. Other elementary particles are built from them; they are their derivatives. Quark-like particles acquire their material forms as soon as their pass over the Planck limit and enter the physical world. Now we may realize that the Planck limit is the marker of the boundary between "real" and "imaginary."

So, a three-layered structure is needed to bring matter into spacetime (see the following diagram). The quantum vacuum or "nothingness" acts as an interface between the world of templates and the quantum world. Therefore, it is not the Higgs boson that is the "God particle." Instead, it is the complex particles that create matter, while the Higgs boson gives mass to the matter particles. The complex particles are the crucial elements of the process of converting "nothingness" into matter. It is these particles that will bring physics closer to …

"complex" reality. It looks like the standard model will have to be rewritten to accommodate this reality.

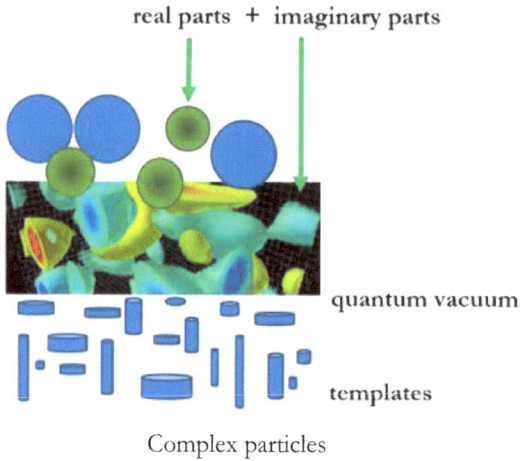

Complex particles

Just like in the case of elementary particles, there are two groups of complex particles. One group includes the complex particles that are needed to create the most elementary forms of matter. They facilitate the "birth" of matter, i.e., convert the invisible "bubbling" fields of "nothingness" into quark-like particles. These complex particles are confined to the quantum vacuum.

The other group of complex particles is needed to mediate the overall shapes of baryonic structures. Atoms, molecules, crystals, planetary systems, and galaxies are examples of such structures. Therefore, other interfaces are needed to facilitate these projections. All interfaces contain imaginary particles. These interfaces play a similar role for the baryonic objects as "nothingness" does for quark-like particles. The difference between them is that "nothingness" facilitates the creation of matter, while the baryonic interfaces enforce the shapes of matter. Therefore, they are the forces which control shapes of physical objects.

What about the "real" components of the baryonic interfaces? After all, the imaginary forms contained within the baryonic interfaces need their fragmentary partners to close the loop between the invisible and the visible. So, what are and where are those fragmentary "real" components that are needed to shape everything in the physical universe?

The answer to this question is surprisingly quite simple.

Every physical object is built from quark-like particles. Now we know that every quark-like particle is inseparable from its imaginary part. This means that every physical object, including atoms, molecules, crystals, planets, galaxies, plants, animals, and humans – contains an assembly of quark-like particles and their imaginary partners. They penetrate the entire structure of every object. We can say that the assembly of imaginary parts constitutes … the imaginary body of every physical object. These imaginary bodies adjust themselves to the forms projected from the world of templates. By adjusting themselves, they carry with them their quark-like partners, i.e., their "real" mates. As a result, the "real" mates enforce the shapes on matter. It is in this way that the shapes of all physical objects are fashioned.

template ➡ [imaginary parts / quarks] ➡ objects

Through these assemblies of quark-like particles and their imaginary partners, all physical objects are linked to the templates within the invisible fields. We are walking within these invisible fields; we are entangled with them.

Occasionally, it is possible to see some "leaks" of these invisible fields around us. Sometimes it is possible to see human-like or animal-like shadows imprinted on a landscape or rocks. For example, the attached photos show such shapes imprinted on 300 million years old stones.

Natural sculptures in la Cote de Granit Rose in Brittany, France
(photos by Dominique Hugon)

Interestingly, physicists have partially identified one such baryonic interface as the mysterious "dark matter." Dark matter is an interface consisting of the imaginary shapes of planetary systems and galaxies. The term "dark matter," however, is not quite adequate for the description of the baryonic interface. Instead, "baryonic vacuum" would be a good term for the interfaces that interact with baryonic structures.

Physicists have chosen the term "dark matter" because they have assumed that its effect on other objects is like the gravitational force. But dark matter is different from gravity. It is much more powerful and sophisticated than gravity. It carries and enforces the overall shapes of planetary systems and galaxies. In this way, it overrides gravity. At this point, it should be mentioned that the classical Newtonian law of gravity is only an approximation of the force which causes large objects to form planetary systems.

Newton's law of gravity is probably the best-known law of physics. It says that the force of gravity between two bodies is proportional to their masses and diminishes as the square of distance between them (see the following illustration).

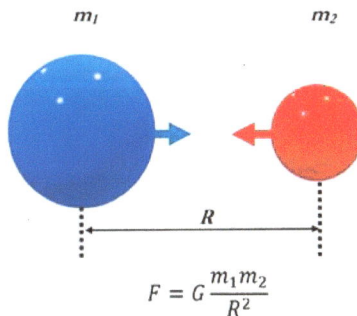

$$F = G\frac{m_1 m_2}{R^2}$$

Newton's law: F is a force, m_1 and m_2 are masses,
R is the distance, and G is the gravitational constant.

Newton's law became widely popularized through an anecdote about a falling apple. Interestingly, this anecdote was first published by Voltaire, Émilie du Châtelet's lover. When Voltaire was in England, he had interviewed Mrs. Conduit, Newton's niece. Apparently, Newton had told Mrs. Conduit that once when he was at his mother's house in Lincolnshire, he had seen an apple fall to the ground. This had made him wonder whether the force that pulled the apple down was something that stretched higher and higher, all the way above the Earth's atmosphere. Voltaire published this story in his *Letters from England*.

Newton's law, however, applies only to a special case of the interactions between objects. Namely, it is limited to the interaction between two objects. A falling apple is a good example of these types of interactions. However, Newton's law is insufficient for the cases where there are more than two objects. The configuration where there are more than two objects is known in physics as the n-body problem. This is a classical problem that covers a large range of situations in astrophysics. For example, the three-body problem applies to the Moon, Earth, and Sun interaction. The inclusion of solar perturbations of the Moon into the interaction between the Sun and the Earth led to the problem's appearance. It was realized that when a third object enters the picture, the situation becomes …. unsolvable. The dynamics of the three-body system is unpredictable; it cannot be solved analytically. Therefore, its behavior may only be described approximately by using elaborate numerical methods. It is even more difficult to resolve the dynamics of the systems which contain four, five, or more objects.

Why is it like this? Why is not Newton's law, the best-known law of physics, sufficient to apply to planetary systems?

The shape of each planetary system is enforced by its corresponding baryonic interface. The interface projects the overall shape of the planetary system which includes a star and all planets orbiting it. The interface provides the complete structure of the entire planetary system. Newton's law, on the other hand, is applicable only to a fragment of the system; it is limited to two selected objects of the system. This is why Newton's law is not sufficient to describe the complete laws that govern the planetary constellations. This also explains why it is impossible to detect gravitons, particles that are supposed to transmit gravity. Now we know that the process of shaping planetary systems is mediated through an assembly of "complex" particles. Through these assemblies of quark-like particles and their imaginary partners, all objects of a planetary system are linked to their corresponding template. Therefore, it is an assembly of "complex" particles that discharges the role assigned by physicists to gravitons.

As indicated earlier, the baryonic interfaces form not only planetary systems but also shape atoms, molecules, and crystals. In these cases, the baryonic interface overrides the electromagnetic force, i.e., it enforces the shapes of atoms, molecules, and crystals.

Today, physicists still believe that dark matter is a physical entity. Therefore, they have focused their efforts on using all types of sophisticated instrumentation to verify their *belief*. For example, the hunt for dark matter is one of the objectives of the James Webb Space Telescope that NASA launched in December 2021. Because of the "complex" nature of dark matter, the currently employed techniques and methodologies are insufficient for this task. A similar technique to the simulation of "nothingness" will have to be implemented to resolve the dark matter question.

It is the first time that scientists are faced with "complex" matter. A new approach must be worked out before modern science comes to terms with this "complex" reality.

148

Entangled Galaxies

I like to experience the universe as one harmonious
whole. Every cell has life. Matter, too, has life; it is
energy solidified. The tree outside is life... The whole
of nature is life... The basic laws of the universe are
simple, but because our senses are limited, we can't
grasp them. There is a pattern in creation.

(*Albert Einstein*)

Scientists detected another region linked to quantum
vacuum. This region is that which is occupied by black holes. As
mentioned earlier, black holes are compared to monsters that
can consume stars, wreck galaxies, and imprison light. In black
holes, matter is converted back into "nothingness."

If we look at the quantum vacuum as the region where
matter is born, black holes are where matter dies. So, the life of
matter goes through a cycle of three stages. In the first stage, the
matter is conceived in the lowest zone of the invisible fields and
delivered onto spacetime in the form of elementary particles.
Then, matter becomes stable and mature when it forms various
baryonic structures. At the next stage, matter is compressed in
black holes and converted back into "nothingness." At this
point, it leaves the physical dimensions. In other words, matter
is born, and then it dies. When it dies, it leaves spacetime.

However, the formation of black holes is not the end of the
process. As mentioned earlier, it was discovered that black holes
gradually leak out some particles and radiation. As a result, black

holes gradually lose their mass and eventually evaporate. The mass that had fallen into a black hole is gradually returned to the universe as particles and radiation. We referred to this effect as the resurrection of matter.

The resurrection of matter brought another problem. The particles swallowed by black holes carry all sorts of information about the various structures they were part of. However, when the content of black holes is converted into "nothingness," all information that it carries is erased. And this is another thing that is inconsistent with the principle of scientific determinism, which states that information must be preserved. Because if any information is destroyed or new information is created, it would be impossible to predict the future or read the past. Any loss or gain would mean that there would be missing information, or some extra information would be gained, so all of physics would collapse. This is why physicists are coming up with various proposals to resolve this problem. One of the proposals often quoted in today's news is the concept of a holographic universe. This proposal assumes that the re-injected particles are the same as that which collapsed into black holes; they are a sort of holographic projection of what fell into black holes. In this way, it is claimed that the information is not lost. Instead, it is gradually returned to the physical universe. So, in this way, the principle of determinism is preserved.

Therefore, one might ask, what is the reason for black holes? Why would nature need such complicated structures as black holes if nothing is changed?

Let's recall that, like in the Chladni patterns, the various shapes of nodes of standing waves may be generated by changing the positions of the points at which the vibrating medium is fixed to its base. Now we may realize that each black hole is such a point. Each black hole is attached or fixed to "nothingness." "Nothingness" is a form of base for spacetime. Therefore, each black hole acts as a fixed point within the vibrating universe.

The locations of these points of attachment are constantly changing. Therefore, this allows for new sets of standing waves to be continuously generated – which determine the overall cosmic structure. In this context, we can look at the map of black holes as an equivalent of three-dimensional Chladni patterns. In this case, spacetime is the oscillating "plate," and black holes are the attachment points (see the attached illustration).

A map of the night sky with supermassive black holes (black holes are marked as white spots; there is not a single star on this map)[36]

36 "New Map Reveals 25,00 Supermassive Black Holes in Night Sky" (https://www.tasnimnews.com/en/news/2022/01/04/2638571/new-map-reveals-25-000-supermassive-black-holes-in-night-sky).

There is another reason for the existence of black holes. Namely, black holes allow for renewing the universe. The new particles re-injected from black holes into the physical universe are not the same as the old ones. By collapsing in a black hole, these particles have been cleansed from their previous history. Therefore, the new particles can form new structures needed at given times and at specific locations. It is in this manner that the universe is continuously evolving. It has been evolving since the very first moment it came into being. It helps to grasp this concept to realize that there are around 400 trillion (40,000,000,000,000,000,000) black holes. Such a quantity of black holes makes quite an effective "cleansing" operation. Or, if expressed in a more illustrative form, this means that a cat ejected from a black hole differs from that which fell into it.

How does this comply with the principle of scientific determinism? It does not.

It does not comply because scientific determinism applies to a closed system. In this case, "closed" means the physical world enclosed within spacetime and the Planck limit. But when matter collapses in black holes, it leaves the physical world and dies. The "dead" matter does not belong to the physical world; it is converted back into "nothingness."

The cleansing of cosmic structures is needed because the entire system is continually updated. It is not just elementary particles that go through adjustments. Massive galactic structures also go through changes. Recent observations indicate that those gigantic galactic structures appear to be synchronized. Namely, their rotations are coordinated despite being separated by billions of light years.[37] This spin-like

---

[37] https://bigthink.com/hard-science/large-scale-structures/?fbclid=IwAR2JnKyl4-YVFJj202oyiaCpL0DUXdNgfMVW58m0e3LcwJC7OH7LEqV8CsI (November 18, 2019).

correlation is striking – as it is equivalent to the entanglement of electrons in an atom. Just like electrons in an atom, these huge clumps of matter seem to be aware of each other. This means that another type of entanglement applies to these gigantic structures. We may call it galactic entanglement.

Template

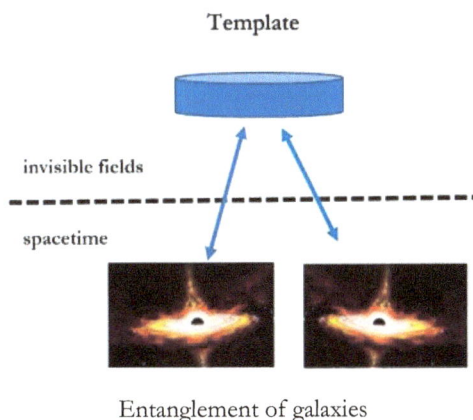

invisible fields

spacetime

Entanglement of galaxies

Galaxies, like electrons in atoms and molecules within quasicrystalline lattices, perform a function requiring them to have a certain degree of awareness of their environment. Again, this kind of awareness is facilitated through a common template which is in a higher zone of the invisible fields. This interface enforces the shapes of galactic clusters.

In this context, it is interesting to compare the cosmos filled in with clusters of galaxies to quasicrystals. Just like quasicrystals, the cosmos appears to consist of structures in the shapes of sheets, filaments, knots, bridges, etc. These galactic clusters are separated by many light years – so there is no possibility for gravitational bounds between them; there is no gravitational interaction between them. This is another indication of the existence of baryonic interfaces, which impose shapes on galactic clusters.

One more thing must be added to have a complete picture of the galactic structures. Namely, the boundary of spacetime is rapidly expanding; it is accelerating. The universe's expansion may be compared to the production of raw material. Expanding spacetime provides more "nothingness" to produce more matter. Matter is the lowest grade of raw material available in the universe. Therefore, the fact that spacetime is expanding – is a sign that the universe is growing; it is still not complete.

The expansion of the universe indicates that there is another interface that projects the shape of spacetime. We may call this interface "galactic vacuum." The operation of the galactic vacuum differs from that identified for baryonic vacuum. As indicated earlier, the baryonic vacuum imposes specific shapes on physical objects, including galaxies and clusters of galaxies. The galactic vacuum, on the other hand, enforces the shape of the entire spacetime. It overrides the baryonic vacuum; it is the most powerful force in the universe. It is the galactic vacuum and not gravity that shapes spacetime.

It may help to grasp the idea of the expanding universe to look at it from another scale. Let's imagine ourselves being placed inside a growing quasicrystal. We would be in almost empty space. Far away above us, we would see clusters of electrons moving seemingly randomly. From this perspective, the process of growing crystal would be incomprehensible to us.

We are in a similar situation when looking at the map of galaxies. It is difficult to recognize the highly sophisticated pattern that all those galaxies are following. It is even more difficult to realize that these remote galaxies form a structure which is needed to preserve living conditions on Earth.

Physicists have partially identified the galactic vacuum. They refer to it as "dark energy" (see the following diagram). Although the term "dark energy" is not accurate because the effect of "dark energy" is very different from that of any known energy.

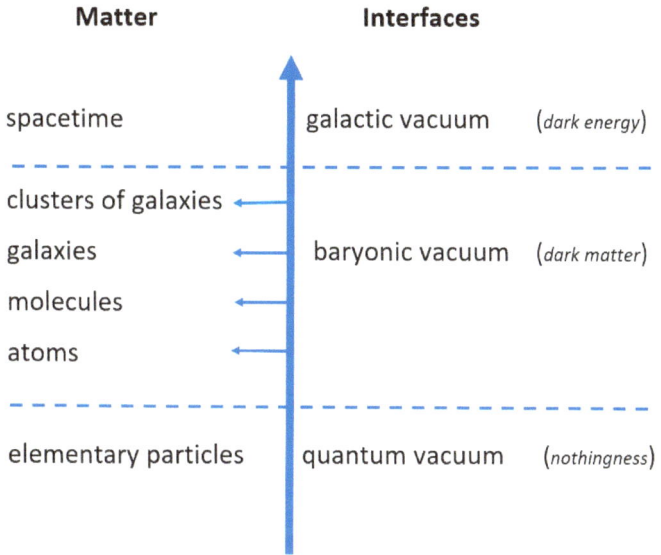

Matter   **Interfaces**

spacetime   galactic vacuum   (*dark energy*)

clusters of galaxies

galaxies   baryonic vacuum   (*dark matter*)

molecules

atoms

elementary particles   quantum vacuum   (*nothingness*)

Matter and its corresponding interfaces
(the currently used terms are in brackets)

Recognition of the baryonic vacuum and the galactic vacuum greatly improves our understanding of matter and the universe. Without them, it will be impossible to construct a "theory of everything," i.e., the ultimate theory of matter.

Now we can see why today's science got into trouble. The deterministic doctrine is based on a bottom-up approach. This means piecing together components to give rise to more sophisticated structures. The concept of Darwinian evolution is based on such an assumption. Nature works the other way around, i.e., it follows a top-down approach, where an overall template is imposed on a structure's components. The world of templates contains the overall shapes of all physical forms projected onto the physical dimensions.

Now it becomes clear that today's science faces quite a challenge. It is the greatest challenge since the controversy between heliocentric and geocentric systems. The deterministic versus perceptive approaches need to be brought together. Just like at the time of Copernicus, it will take some time before the foundation of a new scientific approach is worked out and accepted. Such an adjustment will require loosening up the strictly deterministic view of the physical world.

The Planet Earth

The new worldview that may conceivably grow out of
modern science is likely to be once more geocentric …
(Hannah Arendt)

In the previous chapters we have not yet discussed one kind
of cosmic structure - the planet Earth. So, let's complete the
picture by adding the Earth to the overall design of the universe.

Although the Earth seems to be placed far away from the
center of the universe, it is located within a unique nodal region
of the universe. The nodal region within which the Earth was
formed is the most sophisticated among all the regions of the
universe. The entire universe, with its gigantic clusters of
galaxies, was needed to form and preserve the Earth. The planet
Earth is like a treasure formed within a vast cosmic "ocean" of
vibrating fields. It is like the universe's heart – because it
provides a unique environment within which life can exist. The
Earth is the only place in the entire physical universe where it
was possible to develop and sustain plants, animals, and
humans. It is in this context that the above comment by
Hannah Arendt about "the new geocentric worldview" may be
considered. Hannah Arendt, a German-born American historian
and political philosopher, was one of the most influential
political theorists of the 20th century.

Scientists have been able to work out quite a few details of
the formation of the Earth. According to the big bang model,
the Earth was formed around 4.5 billion years ago,

158

approximately one-third the universe's age. It was made by the accumulation of a cloud of dust left over from the formation of the Sun. Much of the Earth was molten because of collisions with other bodies, which led to extreme volcanism. Over time, it acquired an atmosphere from volcanic outgassing. As the Earth cooled, the lithosphere formed, i.e., the crust and the upper mantle.

The pentangle may also represent the Earth. We may call it the pentangle of the planet Earth.

The pentangle of the Earth
- External star: lithosphere
- First inner star: hydrosphere and atmosphere
- Second inner star: biosphere
- Third inner star: fauna and mankind
- Inner dot: the human mind

In the case of the Earth, a mix of clay, minerals, and water is the equivalent of the grains of sand in the Chladni experiment.

It was this mix which, when placed within a subtle zone of the invisible fields, was transformed into plants, animals, and humans.

In the beginning, the Earth was just like any other planet formed in the universe. There was nothing special about its mineral composition. Then, something started to crystalize. After the lithosphere was formed, oceans appeared, and the hydrosphere took shape. As the next step, a new series of "frequencies" within the invisible fields was activated. These "frequencies" were unique within the universe and were highly localized; they were explicitly focused on the Earth. These frequencies could be activated only within the boundary formed by the hydrosphere and atmosphere. This led to the appearance of biological systems such as plants, flowers, and trees.

Biological systems are much more sophisticated than the previously discussed forms of matter. It is widely accepted that the backbone of a biological system is a very complex molecule known as DNA, which stands for deoxyribonucleic acid. DNA is the information molecule. All living things have DNA within their cells. Nearly every cell in a multicellular organism possesses the complete set of DNAs required for that organism. The DNA is *believed* to contain all the information necessary to build and maintain an organism.

The complexity of DNA is schematically illustrated in the following diagram. DNA is made of two linked strands that wind around each other to resemble a twisted ladder – a shape known as a double helix. DNA's double helix has become one of modern science's most well-known and iconic images.

The information in DNA is stored as a code made up of four chemical bases, known as adenine (*A*), guanine (*G*), cytosine (*C*), and thymine (*T*). DNA bases pair up with each other, i.e., *A* with *T* and *G* with *C*. Each pair of bases is attached to two long strands made of a sugar molecule and a phosphate molecule. The code is made from the vast number of available combinations of the DNA's bases.

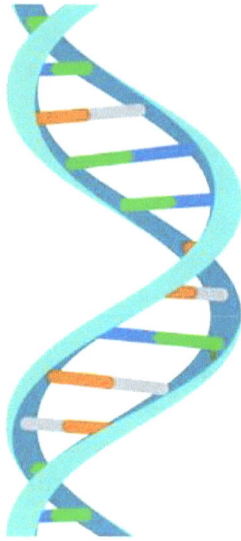

DNA's double helix[38]
The two long-twisted strands are made of a sugar molecule and a phosphate molecule. The four chemical bases which join these two strands are adenine (*green*), guanine (*orange*), cytosine (*grey*), and thymine (*blue*)

The discovery of DNA's double-helical structure in the 1950s is the most significant accomplishment in biology of the 20th century. Knowledge of this remarkably sophisticated structure provided critical insights into how DNA could serve as the information molecule of all living systems. DNA also serves as the primary element of heredity in organisms.

---

[38] https://www.genome.gov/genetics-glossary/Double-Helix

Whenever organisms reproduce, a portion of their DNA is passed along to their offspring. This transmission of all or part of an organism's DNA helps ensure continuity from one generation to the next while allowing for slight changes that contribute to the diversity of life.

This means that in the case of biological systems, it is not an atom or a simple molecule that serves as the fundamental building block. Instead, the basic building block is in the form of DNA, a complex assembly of molecules. The way DNA functions is often compared to the way in which letters of the alphabet appear in a certain sequence to form words and sentences. But the main point of this comparison is that the formation of words and sentences requires a ... dictionary. Because only a specific sequence of letters can make it into a meaningful word; it is not any random collection of letters. Otherwise, we would end up with Jorge Borges' *Library of Babel*, i.e., shelves full of books written by a random permutation of letters. Therefore, an overriding template must serve as a "dictionary" for assembling DNAs. This means that the multiplication of DNA that leads to the formation of living organisms must also be ... a non-local process.

Now, we can recognize a certain similarity between the growth of quasicrystals and the way DNA forms living organisms. Of course, DNA is much more sophisticated. But there is an interesting observation that can be inferred from this comparison. The investigation of quasicrystals showed that their assembly cannot be achieved by simply adding atoms one at a time. Instead, it needs a link to a template in the invisible fields, which acts as a "dictionary." This means that the process of growing quasicrystals is necessarily non-local. Therefore, the same condition must also apply to biological systems. Just like electrons in atoms and molecules in quasicrystals, DNA molecules follow the design that is projected from the invisible world. DNAs are entangled through their common links to a corresponding template within a baryonic vacuum. Otherwise, it would be impossible to explain the process satisfactorily.

Just like the classical model of crystal growth, the currently accepted process of the growth of biological systems is incomplete. The biological growth based on DNA is only a part of the process. The other part is still to be worked out.

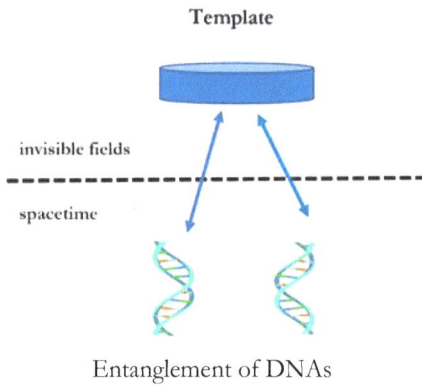

Entanglement of DNAs

Now we may realize that the templates, stored in the invisible fields, function as the actual code carriers for every physical object and every biological system. These templates serve the purpose that scientists have assigned to DNA. This is another conceptual barrier that must be overcome to advance our understanding of the universe.

***

Plants, weeds, and grass covered the surface of the lithosphere, on land and at the bottom of oceans. The flora formed a membrane that would act as a boundary within which higher forms of life would be activated. We can call a space within that boundary "biosphere" – although, in this context, the term has a different meaning than that commonly used. Namely, the biosphere provided a medium within which birds and creatures living in oceans and on land could be formed.

Scientists have realized relatively recently that the flora is much more sophisticated than previously thought. For example, it has been found that trees form quite advanced underground communication networks. These networks are created by fungi joining with plant roots. Trees use their networks to do such things as communicate and share resources. Some scientists call it the internet of trees, or the "wood wide web":

> The trees, understory plants, fungi and microbes in a forest are so thoroughly connected, communicative and codependent that some scientists have described them as superorganisms. Recent research suggests that mycorrhizal networks also perfuse prairies, grasslands, chaparral, and Arctic tundra – essentially everywhere there is life on land. Together, these symbiotic partners knit Earth's soils into contiguous living networks of unfathomable scale and complexity.[39]

The surprising thing about these communication networks is their purpose. It was found that the main reason for the existence of these underground communication networks was quite altruistic. Namely, trees use these networks to help one another at their own expense! For example, resources are sent from the oldest and biggest trees to the youngest and smallest.

---

[39] "The Social Life of Forests" by Ferris Jabr, *The New York Times*, December 2, 2020.

And if a tree is on the brink of death, it sometimes donates a share of its carbon to its neighbors. This sort of altruism contradicts the core tenets of Darwinian evolution. Since Darwin, biologists have believed in the struggle of each organism to survive and reproduce within a given population. Underlying all of these was the single-minded ambition of selfish genes. Consequently, trees were regarded as solitary individuals that competed for space and resources and were otherwise indifferent to one another. And now, it turns out that the overall modus operandi of trees has a greater focus on cooperation over self-interest and on the emergent properties of living communities rather than individuals.

Trees are one of the biggest elements of the biospheric boundary. Therefore, the altruistic behavior of trees should not be a surprise. The function of this boundary is to provide a medium within which quite sophisticated nodes could be generated and sustained. Therefore, in this case, it is not so much altruism but something that must be done. Something that is written into the original plan.

Looking at the migration routes of birds and whales may help us understand the biosphere's function.

For example, a bird called the Arctic tern undertakes the largest migration of any bird in the world. It travels all the way from the Arctic to the Antarctic and back again each year. Recent studies have shown that its average annual round-trip length is about 72,000 kilometers. The long journey ensures this bird sees two summers per year and more daylight than any other creature. In old times, this species greatly interested mariners, who took their bearing from its flight path.

The interesting thing is that their breeding and nesting take place during two summers in the same year. One summer in the north, the other summer in the south. But why do these birds need to fly from summer in the northern hemisphere to summer in the southern hemisphere?

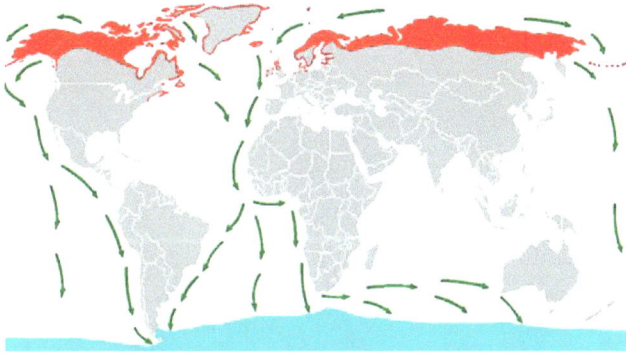

Migration routes of the Arctic tern[40]
(breeding grounds are indicated in red; nesting grounds are marked in blue)

As far as height is concerned, most birds fly relatively low, i.e., under 150 meters. During migration, however, many species fly at 600 to 1,500 meters or higher. The record height is held by the Rüppell's griffon. This vulture is found in the Sahel region of central Africa. This highest-flying bird has been recorded to fly at an altitude of 11,000 meters. But why do these birds have to fly so high? The most common answer is that this allows them to get away from predators. Another explanation is that it does help them to hunt.

Among the creatures living in the oceans, whales are accomplished divers. Their deepest recorded dive was nearly 300 meters. Whales also undertake some of the longest migrations. They are often swimming up to 20,000 kilometers

[40] https://en.wikipedia.org/wiki/Arctic_tern#/media/File:Sterna_paradisaea_distribution_and_migration_map.png

166

every year. Researchers have not yet established why whales put so much energy into their migration effort. Definitively, whales do not migrate because of looking for appropriate sites to give birth to their calves. Because of their size, large whales should be able to give birth in frigid polar waters. Neither does food seem to be the reason for their migration. Quite the contrary, due to reduced feeding opportunities in the tropics, most whales fast during their months-long migrations. So why go to the trouble?

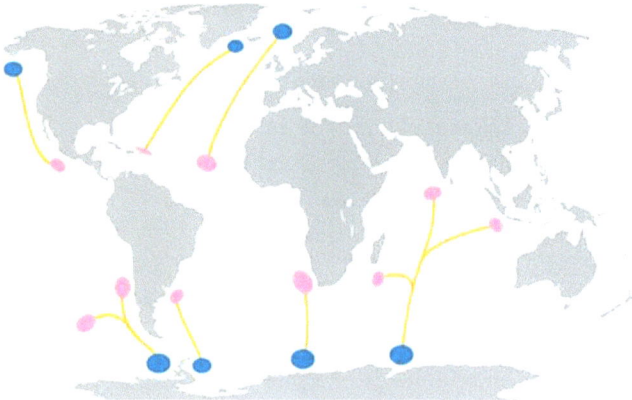

Migratory routes of the blue whale[41]
(feeding grounds are marked in blue; breeding grounds are in pink)

We can find the answers to these questions by looking at the shapes of the migration routes of the whales and the Arctic tern. This picture will be completed if we present it in three dimensions by adding the altitudes at which the various birds fly and the depths of whale's dives. If we discard the contours of the continents, then we may be able to recognize the patterns of these migration routes. It should come as no surprise that the shapes of the migration routes look very much like ... Chladni

[41] https://www.offthemap.travel/news/understanding-whale-migration/

patterns. Yes, that's correct. The migrations routes of these species correspond to the distribution of nodes induced by the invisible fields within the boundary of the biosphere.

Now we can understand why these various species follow their seemingly irrational routes. Namely, they simply follow the patterns formed within the biosphere. It is not the search for food nor nesting sites that determines the migration routes. It is the other way around. The routes are the prime factors; food and nesting are the secondary aspects:

> What use is the wind to an eagle or the ocean to a dolphin? … They soar and spin and swoop while we put one foot after the other. When the air or the water moves over something from one edge of the world to another, they know.[42]

There is another important aspect to the migration routes. Namely, the migration routes of whales and birds mark the boundary of the biosphere. The continents are like a cross-section of the biosphere. In other words, the biosphere provides an enclosure within which more subtle nodes could be projected. These more subtle nodes within the biosphere led to the appearance of animals. Afterwards, groups of human nomads appeared.

It may help to grasp this concept to recall the three-dimensional shapes of the Chladni patterns formed by a relatively simple acoustic field. Therefore, it should be no surprise that more subtle fields, like the invisible fields, can create more sophisticated structures. Just like grains of sand can

---

42 *The Mines of Light*, Arif Shah, Pick and Shovel, LLC, Los Angeles, 2016.

form very complex patterns, so within the biosphere – minerals can form the shapes of humans:

> Man and the earth are made of the same things. ...
> Water and salt and minerals and elements,
> all mixed up and held together
> by this thing you call your mind.[43]

Mankind is placed on the top of the biological structures. The previously mentioned observation about the non-locality of the formation process of living organisms also applies to humans. Just like in the case of other systems, the shape of humans is a projection of a template from the invisible world. This means that humans are three-dimensional nodes that walk within the inner space of the biosphere. Like all other earthly creatures, humans are formed from the elements which are available on the Earth. Humans appeared in that most subtle region of the biosphere. We can call this space the human-sphere.

It is important to notice that there is a functional distinction between flora and higher forms of life, such as fauna and humans. Flora forms the boundary of the biosphere. The boundary determines the region needed for the formation of a series of nodes within the volume of the biosphere. These nodes take on the shapes of animals and humans. This is why sustenance of the biosphere is so critical to the overall purpose of the cosmic plan. In this plan, humans are the last link needed to complete the plan. Humans are the most precious element of the entire enterprise.

---

[43] Ibid.

***

As we may see, the function of the flora is much more sophisticated and fundamental than is usually considered. The various species' shapes, locations, and arrangements are of great importance. And this is a pointer for those interested in the preservation of the natural environment of the planet. It is not only beauty, green spaces, and clean air that are at stake. These are secondary factors. The primary factor is the preservation of the boundary of the biosphere, so the nodal structure that forms and sustains life on this planet – remains in sync with the galactic dynamics of the universe.

170

Are We Alone in the Universe?

The silence of the night sky is golden
… in the search for extraterrestrial life;
no news is good news.
It promises a potentially great future for humanity.

(*Nick Bostrom*)

Are we alone in the universe? This is one of the most important questions that science has been trying to answer for us.

One of the prime objectives of space research is to look for traces of life outside of our planet. The entire space research is based on an unspoken assumption that out there are some traces of life. It is only a matter of time and technology to find some. Therefore, the search for life outside the planet is seemingly fully justified. Scientists feel compelled to search for life, regardless of how enduring such a task would be, even if it takes … forever.

We are bombarded in daily news feeds with new discoveries which indicate either the existence of a new type of exoplanet, or the presence of water and some earthly minerals on some remote planets, or the signs of past life in various rocks on Mars. And this leads to the further justification of the efforts to pursue these various hypothetical indications. Because it is believed that any evidence that there was once life somewhere in outer space – would confirm that life is abundant somewhere else in the cosmos.

In the context of the material presented in this book, such an assumption about life outside of the planet Earth does not fit into the overall scheme of things. Therefore, looking for traces of life outside our planet will not help advance our knowledge of the Earth, the universe, or life itself. Quite the contrary, it will significantly delay it. It will slow the progress of knowledge because it is based on a false assumption. It would be much more effective to rephrase the goal of the entire space enterprise by asking the question: Why is there no life outside of the planet Earth? Because there is no life out there.

Space is somewhat like a vast desert. It is filled with flying rocks of various sizes, hot and cold gases – and that's it. As has been pointed out, in space - matter is born and then dies. It is a spectacular process, full of light, smoke, gases, liquid rocks, and incredibly complicated structures. But the inner fabric that is needed to give rise to life and sustain it – is absent.

In this context, listening to people who had a chance to fly into space may be interesting. For example, William Shatner, a Canadian actor who played Captain James T Kirk in *Star Trek*, was given the opportunity to go to space in October 2021. After decades of playing a science-fiction character who was exploring the universe and building connections with many diverse life forms and cultures, he thought that he would experience a similar feeling: a feeling of deep connection with that immensity around us, a deep call for endless exploration. A call to go boldly where no one had gone before. Here is his reflection about his actual space experience:

> While I was looking away from Earth, and turned
> towards the rest of the universe, I didn't feel connection;
> I didn't feel attraction. What I understood, in the clearest
> possible way, was that we were living on a tiny oasis of

life, surrounded by an immensity of death. I didn't see infinite possibilities of worlds to explore, of adventures to have, or living creatures to connect with. I saw the deepest darkness I could have ever imagined, contrasting starkly with the welcoming warmth of our nurturing home planet.[44]

One does not have to go to space to have a similar experience. For example, it is enough, and it is much less expensive, to take a hike along the volcanic lava fields on the slopes of the Etna in Sicily.

On the slopes of the volcanic fields of the Etna, Sicily
(photo credit: Dominique Hugon)

When in the middle of those fields, surrounded by the black mass of lifeless molten rocks – one may feel exactly like Captain James T Kirk did. It is a very disheartening feeling, very

[44] https://www.theguardian.com/environment/2022/dec/07/william-shatner-earth-must-live-long-and-prosper-aoe?CMP=Share_AndroidApp_Other

uncomfortable. Something is missing, something vital – although it is hard to define at first what it is. Then, one realizes what it is if one looks further away from those black surroundings towards the horizon, where it is possible to see trees and lush greenery. Over there is … "home." These molten rocks made a "black hole" within the biosphere. In the history of the Earth, this small chunk of land belongs to the time before the formation of the biosphere; before the planet became our "home." When we are in the middle of the lava field, we are partially outside of our natural nodes, just like Captain James T Kirk described in the above-quoted reflection.

What about aliens or UFOs? Are these not indications of life outside of the planet Earth?

We can skip aliens and alien civilizations. These, we know, do not fit into the overall picture. Therefore, there is no possibility of their existence.

UFOs (unidentified flying objects) are a bit of a different phenomenon. There is always a possibility within this huge dynamic universe floating within the invisible fields that some sporadic leaks from the invisible to the visible may occur. We may expect that such leaks would take the forms of very unusual shapes and colors. They could be easily discerned from other terrestrial phenomena.

According to Heather Dixon, the head of national investigations of the British UFO Research Association (BUFORA), only two percent of reported cases of supposed UFOs could not be explained.[45] These two percent could be related to the events mentioned above. All the rest can be easily

---

[45] "Most UFOs - like the Chinese spy balloon - can be explained away. But what about the other 2 percent" by Heather Dixon, *The Guardian*, Feb 16th, 2023.

explained. Here is a list of the most common effects, which have been many times reported as UFOs:

- Meteor fireballs: to the naked eye it is a yellowish object which seemingly appears out of nowhere, flying fast and silently across the sky and leaving a glowing trail behind. Then, it suddenly breaks into smaller pieces before vanishing into thin air – all in under a minute.
- Lens flare: effect related to the light which sometimes bounces off the lens in a camera, goggles, glasses, or windows, causing a lens flare. Some flares can look like solid objects accidentally framed within the field of view. These are often mistaken for unworldly flying objects.
- The International Space Station: this is a vast platform, larger than a football field. It can be significantly brighter than most night sky objects. It moves fast, taking a few minutes to cross the sky from one horizon to the other.
- Constellations of military satellites: some of the surveillance satellites consist of a trio of satellites that orbit in a triangle formation and are sometimes visible to the naked eye.

Nevertheless, many research institutions have seriously considered the possibility of UFOs and alien civilizations. This is because the psychological impact of the existence of "others" is so strong that it cannot be easily erased from the human imagination. Apparently, humans do not respond well to "otherness." Therefore, some scientists embarked on a project to prepare us psychologically for an encounter with alien beings. This approach went even as far as to use robots to facilitate such a hypothetical interaction.

Nick Bostrom, the above-quoted Swedish-born philosopher at the University of Cambridge, recognized that no evidence of extraterrestrial life "is good news." Because any evidence of previous life in other parts of the cosmos would indicate that life on Earth would end in the same way. Therefore, the lack of such evidence "promises a potentially great future for humanity."

So, let's leave the subject of UFOs and alien civilizations here. There are more important things to consider.

How to Enhance One's Perceptiveness

All the answers are on the board.
(*Arif Ali-Shah*)

As we have seen, the quantum world operates with strange and fanciful rules. It turns out that some of those rules have found their parallels in a recently introduced chess-like game. And this game may help to develop the skills needed to advance perceptive science. So, let's take a closer look at the overall design of the game, which has some elements of quantum mechanics.

First, let's start with traditional chess.

The currently known form of chess emerged in Southern Europe during the second half of the 15th century. Its origin, however, may be traced to a much older game known as chaturanga, which originated in India at the time of the Gupta Empire around the 6th century. Today, chess is one of the world's most popular games, played by millions of people worldwide.

Chess is played by two players on a square chessboard arranged in an eight-by-eight grid. One player plays with white pieces and the other with black pieces. At the beginning of the game, each player controls sixteen pieces: one king, one queen, two rooks, two bishops, two knights, and eight pawns.

A traditional chess board

The game's objective is to checkmate the opponent's king, whereby the king is under immediate attack (in "check"), and there is no way for it to escape. It is generally assumed that chess is a strategy game with perfect information – for each player can always see all the pieces on the board.

Chess is a rather complex game because of the high number of possibilities. The complexity of chess was not precisely defined until 1950. In that year, the American mathematician Claude Shannon wrote in *Philosophical Magazine* an article entitled "Programming a Computer for Playing Chess." Shannon wondered whether it was possible to construct a machine to play a perfect game of chess. As the starting point, he calculated the number of all variants of the game. He assumed that each of the two players might choose one out of 30 possible moves (plies) in each round of the game. A round consists of two consecutive plies, i.e., one by whites and then by blacks. This means that there are 30 x 30 (= 10^3, approx.) possible variants in each round. A typical game lasts about 40 rounds. Therefore, the estimated number of variants for a typical game is in the range of $(10^3)^{40} = 10^{120}$. This estimate is known as the Shannon number. This is an incredibly big number, i.e., 10 followed by 120 zeros.

It helps to get a sense of the magnitude of this number to consider the largest possible value that has a physical meaning. Such a value would correspond to the ratio of the largest object to the smallest one that exist in the universe. The smallest number which has a physical meaning is 1.6 x 10^{-35} meter, i.e., the Planck length. Now, let's construct a cube with each side equal to the Planck length. Such a cube is the smallest three-dimensional object that could exist; nothing in the physical world can be smaller than that. Therefore, this cube may serve as a basic unit to measure volume; let's call it the Planck cube. The entire physical universe is on the other side of the scale. This is the largest entity that exists in spacetime. So, if we divide the volume of the universe by the Plank cube, we will get the largest possible number that has physical relevance. The radius of the universe is approximately equal to 10^{26} meters. Knowing the radius makes it possible to estimate the volume of the universe and express it in Planck cubes. The result is 10^{184} Planck cubes (i.e., 10 followed by 184 zeros). This is the number of Plank cubes needed to fit into the entire volume of the universe. We can refer to this number as the Planck number. This is the largest number that has physical meaning. Anything larger than that – is just abstract.

Now, let's go back to chess. We can compare the Planck number with the Shannon number. It turns out that the Shannon number is 10 vigintillions (10 x 10^{63}) times smaller than the Plank number. In this context, the Shannon number is still within the physical domain. As soon as the Shannon number was known, it was possible to start work on a chess-playing machine. But it wasn't until the 1970s that computers began to defeat humans in the game. In the beginning, the computers played against ordinary players. Those early chess computers were no match for Grand Masters.

It took two more decades to demonstrate that the chess game was strictly deterministic in its very nature. In the mid-1990s, a chess-playing expert system was developed and then installed on a unique purpose-built IBM supercomputer. The system was named Deep Blue. It first played against the reigning world champion Gary Kasparov in a six-game match in 1996. Deep Blue lost two games to four. The next year, however, an upgraded version of Deep Blue defeated Kasparov by winning three and drawing one. Deep Blue's victory is considered a milestone in the history of artificial intelligence. With Deep Blue's victory, chess lost some of its mystery and attractiveness.

The development of various chess-playing machines followed Deep Blue's victory. Computers have the ability to analyze the totality of the game in a way that dramatically outperforms humans. The best human chess players spend thousands of hours researching previous chess games and theorizing new lines of play. The thing is that modern chess engines have become so powerful and widely available that even the world's best players don't stand a chance against a software that anybody can now download free of charge.

The arrival of chess software demonstrated that traditional chess is simply a binary game. It is based on a relatively simple mode of operation, which may be summarized as "either – or." In other words, binary chess reflects an automatic type of operation of the human mind. It has been observed that this type of operation dominates most human activities. Namely, when faced with almost any situation, the human mind decides whether to accept or reject it. Its weakness is that, if it becomes the only mode of approach to a situation, it effectively screens the individual from another kind of perception, which - as we will see later - is essential for the sustenance of humanity.

***

It was probably no coincidence that, at one point, a new type of game would appear, requiring a more advanced operation of the human mind. Such a game was introduced in 2014. This new game is called quaternity. It may help to deal with such complex issues as those encountered by today's physics.

Arrangement of pieces on the quaternity board[46]

---

[46] All the quaternity screenshots are taken from the *Quaternity*™ online platform at https://play.quaternity.com. They are published with permission from Quaternity LLC. All right reserved 2023.

In quaternity, instead of two – there are four players.[47] Each of the players is playing against the other three. And this brings the game into entirely new territory. Instead of the usual eight-by-eight pattern, the quaternity board is checkered with twelve-by-twelve white and black squares. Each of the four sets of white, red, black, and green pieces is identical to those used in regular chess, i.e., king, queen, two bishops, two knights, two rooks, and eight pawns. Instead of the usual two rows on opposite sides, the pieces are arranged in the corners of the board. Each set of pieces is contained within a five-by-five squares territory.

The kings are positioned in the corners. In front of the king, on the diagonal, is the queen. Three pawns occupy the left front line of each set; three other pawns are on the right front line. These six pawns can move along their respective files (columns) and ranks (rows). They are called "committed" pawns. In addition, there are two "central" pawns. Like the king and the queen, the central pawns are positioned on the main diagonal of the board. One of the central pawns joins together the two front lines of the committed pawns. In front of it is the advanced central pawn. One of the rooks and two knights are in the row just behind the three pawns on the left. In the row behind the pawns on the right are two bishops and the other rook. The overall arrangement is a fascinating geometrical combination of colors and patterns.

Except for the central pawns, the moves of the pieces are identical to those used in regular chess.

---

[47] The rules of the game are described in the following books:
Beginners Guide to Quaternity by Javier Romano, London, 2022;
Quaternity – Nuestro método (in Spanish) by Jorge Mas Sirvent and Jorge Yago Ferreyra, Editorial SUFI, Madrid, 2022.

Moves of the quaternity pieces:
- Kings move one square in any direction, as long as that square is not attacked by the pieces of the other players (there is no castling allowed).
- Queens move diagonally, horizontally, or vertically any number of squares.
- Rooks move horizontally or vertically any number of squares.
- Bishops move diagonally any number of squares.
- Knights move in an 'L' shape': two squares in a horizontal or vertical direction, then one square vertically or horizontally. They are the only pieces able to jump over other pieces.
- Pawns move forward one square along their rows/columns. However, the two central pawns may make their first move either vertically or horizontally. After making the first move, the central pawns become "committed" to the direction of their first move.

The board's design and the pieces' initial arrangement drastically increase the game's complexity. Let's estimate the number of possible game variants so we can compare it with the Shannon number calculated for traditional chess. We may follow Shannon's calculation method to determine the number of available variants of the quaternity game.

The first thing to do is to find out how many moves are possible for each player in a ply (a single move by a player). Each of the four players can choose one out of 41 allowed moves in the first ply. This makes 2.8 million variants for the first round. In this case, a round consists of four consecutive plies by the four players. As the game progresses, more squares become open to bishops, rooks, and queens. Therefore, the number of variants can go up to 4.8 million per round. For example, the following diagram illustrates a possible arrangement on the board in the middle of a game (at the end of the 30th round, i.e., after 120 plies). At this point, there are 4.7 million variants available for the next round.

Arrangement of the pieces after 30 rounds (i.e., after 120 plies)

During most of the game, the number of variants per round fluctuates between 2.8 and 4.7 million. It is only in the final stage of the game that the number of possible variants drops. On average, there are about 3 million possible variants per round. Usually, it takes 200 – 280 plies to finish a game. This corresponds to 50 – 70 rounds.

This means that the number of available variants in a 50-round game is in the range of $(3 \times 10^6)^{50}$. If you enter this formula into a scientific calculator - you will get the answer: infinity! Of course, a larger computer can give the numerical answer which is in the range of 10^{300}. In practice, however, such an answer is meaningless. The bottom line is that the number of possible variants in the game of quaternity, i.e., the Q number, is more than googol times higher than the Planck number.[48] This means that the number of possible variants is outside the physical realm.

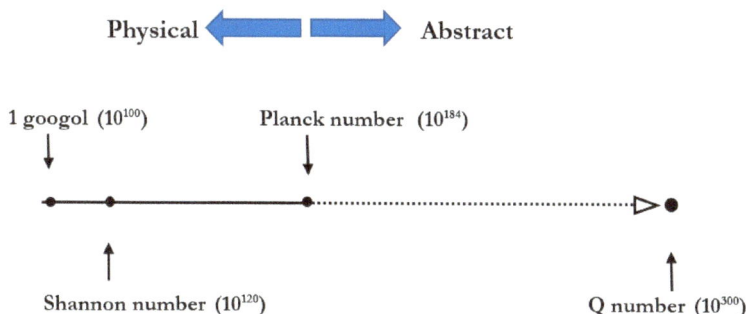

The relation between Shannon and Q numbers

The complexity of quaternity, however, is not only related to the number of possible variants of the game. The other factor that contributes further to the game's complexity is that there are four players playing against each other.

When playing quaternity, one is playing against three opponents on the same board at the same time. So, there are six simultaneous games being played between the four players. In other words, each player must follow and respond to all these six elements of the game. This situation is equivalent to the *n*-body problem, which was described previously. In this context, one may look at traditional chess as an analogy of the two-body problem. Traditional chess is a deterministic interaction between two players. This is why it is possible to have a computer compete effectively with humans.

---

[48] A googol is the number 10^{100}. In decimal notation, it is written as the digit 1 followed by one hundred zeroes. The term was coined in 1920 by 9-year-old Milton Sirotta, a nephew of U.S. mathematician Edward Kasner. Milton may have been inspired by the contemporary comic strip character *Barney Google*.

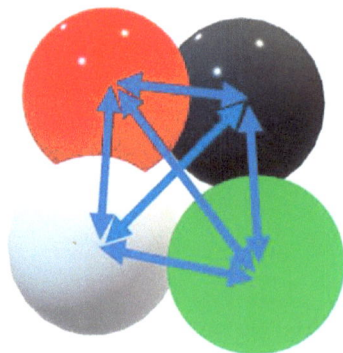

Quaternity as a four-body problem

Quaternity, on the other hand, may be looked at as a special case of the four-body system, i.e., each of the four players interacts with three other players. These mutual interactions are like extra dimensions that add further complexity to the game. Therefore, the game is outside of the rules of probability and calculations. It is impossible to have a binary computer capable of playing a perfect game of quaternity. Quaternity provides an entirely new medium for exercising those modes of operation of the mind that humans possess but which remain in their latent states. We can refer to these latent states as various degrees of perceptivity.

Playing quaternity is like experiencing a fairy tale. However, you are not only reading it. Instead, you are writing your own tale, and as you are writing, you are living it. And your tale has a unique structure, which provides a degree of flexibility but within certain boundaries. By experiencing these self-made tales, one may start to perceive a sequence of events from an entirely new perspective. Exposure to these sorts of experiences brings into action the right hemisphere and, at the same time, attenuates the left hemisphere. In other words, it stimulates the brain to switch from a single mode – into a simultaneous operation of sequential and holistic modes.

It may help further to discover the game's richness to look at the quaternity board as a certain variant of the Chladni patterns. The board is like a plate placed within a certain field. This field induces waves, which can propagate along specific patterns.

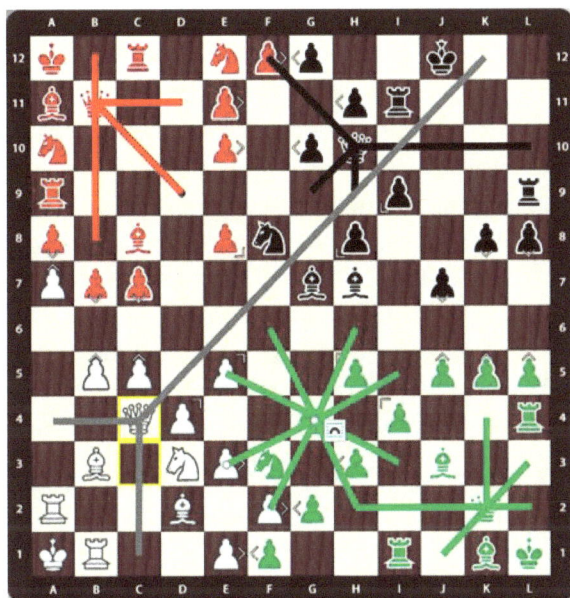

Standing waves of the four queens and the green knight

The patterns are determined by the geometry of the board and the "natural frequencies" of its vibrations. In this representation, each piece may be seen as a node of a standing wave which propagates along specific squares of the board. Or, if expressed in the context of quantum mechanics, each piece on the board acts as a particle wave: both ends of its standing wave are attached to a piece or an edge of the board.

The above image illustrates an example of such standing waves for the four queens and the green knight (G4).

The lines mark those squares of the board that the queens can travel through and the squares that the green knight can move to. One can recognize the knight's paths because they are neither parallel nor perpendicular to the board's diagonals; furthermore, the length of each knight's path is always the same.

If all the pieces on the board are treated as nodes of such waves, then one could have a very complex Chladni-like arrangement of a multiplicity of standing waves. To look at those patterns without the board's background may help to realize the relevance of such a representation. For example, the graphical representation of the four queens and the green knight is shown in the following diagram.

At first look, this graphical representation without the board and the chess pieces' images may seem meaningless. Yet, one can easily recognize which pieces are represented by those various figures and where they are positioned with respect to the board. Such an illustration contains a very precise record of a particular moment of the game. Therefore, it is possible to use such patterns to represent and analyze the game. As in the case of the Chladni patterns, these graphical figures may also be looked at as sort of … musical notes. Although the sounds they represent belong to "music that may not be heard," i.e., it is not ordinary music.

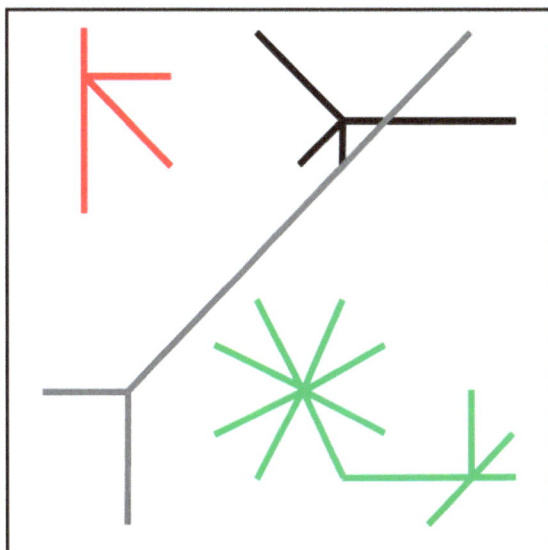

Q-patterns of the four queens and the green knight

In this presentation, the board is placed within the quaternity field. This field spreads throughout the entire board. The field is invisible on the board, but its presence and characteristics can be inferred from the "behavior" of the pieces. Consequently, every piece on the board is like a ripple of the underlying quaternity field. It is in this context that the Q-patterns may help to unravel some nuances of quantum mechanics.

***

The aim of the game is to checkmate the opponents' kings. The checkmates are the critical elements of quaternity. As there are three opponents, therefore the game leads through three consecutive checkmates.

One can use the Q-patterns to illustrate the design and the execution of a checkmate. The following two illustrations show a sequence of moves leading to a checkmate. The players make their moves by taking turns in a clockwise direction, i.e., starting with the whites, followed by the reds, the blacks, and then the greens. The sequence illustrated here starts with the player playing reds moving his queen from square D9 to square L1. In this way, the red queen puts the green king (H1) in check:

A checkmate sequence (the red queen checks the green king)

The green king is checked but is not checkmated. The green king has several options to get out of check. For example, the check can be temporarily blocked either by the green rook (to J1) or by the green bishop (to K1). Another option is to find a safe square within the so-called royal squares, i.e., the squares which surround the king. There are five royal squares in this situation: G1, G2, H2, I2, and I1. However, the squares G1 and

I1 are under the red queen's attack. The green pawn occupies square G2. The green king cannot move to square H2 – because that square is under the white pawn's attack. Therefore, there is only one safe square – I2. Because of these options, the green king is not checkmated yet. However, before the green player can move his king and escape from check, he must wait for the black player's move.

The black player can also check the green king by moving his bishop from H7 to K4 (see the following diagram).

A checkmate sequence: the black bishop
completes the checkmate of the green king

Now the green king is double-checked by the red queen and by the black bishop. There is no way to block both checks. Neither can the green king escape to the square I2 – because this square is now under the attack of the black bishop. In this way, the green king is checkmated.

As we can see, three players' pieces were needed to complete the checkmate: the white pawn, the red queen, and the black bishop. However, the credit for the checkmate goes to the player playing the black pieces – because the black bishop completed the checkmate. In accordance with the game's rules, all remaining green pieces undergo a sort of transformation. Namely, the green king disappears, and all green pieces change their color to black, the color of the checkmating piece:

Post-checkmate transformation of the green pieces

At this point, it is interesting to quote Richard Feynman, a great American physicist who received the Nobel Prize in 1965 for the development of quantum electrodynamics. Feynman compared Nature and the way it operates to a sort of game of chess being played by God. And what physicists are trying to do, according to his analogy, is to figure out the rules of that game. They do not know the game – but are able to observe the movements of things, just like observing the movement of pieces on the board. Feynman alluded to the possibility of the next revolution in physics – which would take place when, one day, a new rule of the game is discovered. He suggested, as an example, such a rule as a bishop changing its color. Interestingly enough, such a rule is part of quaternity. As shown in the above diagram, when one of the players is checkmated, all his pieces (not just a bishop) change their color. According to Feynman's prediction, such a new rule would mark a coming revolution in physics.

So, what is this critical feature that is hidden behind the changing of a bishop's color that would mark a revolution in physics? What kind of a new property of matter does it imply?

In terms of quaternity, changing the bishop's color is the result of a checkmate. Is it then possible to find a quantum effect analogous to a checkmate?

***

In quantum measurements, an experimenter must consider all the possible options to capture or "measure" a particle. When captured, an electron or a photon is transformed from its wavy form into a particle. As a result, the particle ceases to be a wave. Technically, such capture is called a wave function's collapse. Because it is then that the probabilistic nature of the wave is transformed into a particle-like form: a wave of probability is turned into a fact. The collapse is one of the mysterious aspects of the quantum measurement problem.

194

Let's illustrate the sequence leading to the previous checkmate by using the Q-diagram. The overall layout of the scheme of the checkmate is presented below. The green dot indicates the position of the green king. The arrowheads (red, grey, and black) and the green square indicate all the five "royal squares" surrounding the green king.

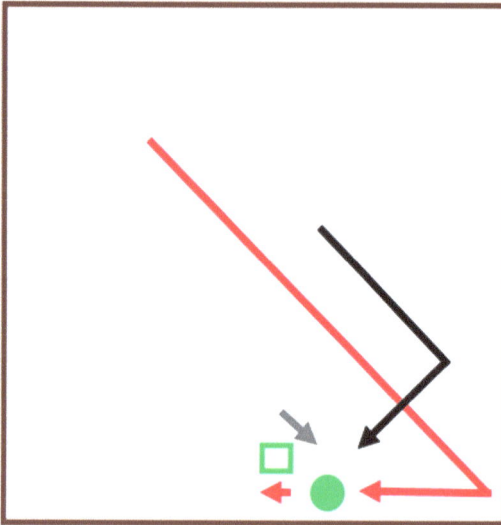

Q-pattern of the checkmate

According to the Q-diagram, the red, grey, and the black lines are wave-like oscillations propagating along their allowed paths. In this representation, the result of the checkmate is that the green king has been confined in such a way that it cannot "oscillate" anymore; it ceases to exist as a "wave." When checkmated, the king's wave function collapses; the king's state is transformed. We could say that the king has been "measured;" the king ceases to exist as a wave. This indicates that it is not an act of passive observation that transforms the wave function into a particle. Instead, it is an act of "checkmating" or confining a particle in such a way, so it is not able to oscillate anymore. In other words, the mysteriousness of the quantum measurement is equivalent to "checkmating" a particle!

But this is only one of the parallels between quaternity and quantum mechanics.

As we have seen, a set of pieces goes through a dramatic transformation when checkmated. The checkmated king disappears, and the checkmated pieces change their color. This means that only three sets of pieces are left on the board after the first checkmate. This type of transformation continues with each of the two following checkmates. An example of the situation after the second checkmate is illustrated here:

Second transformation
(after the second checkmate; only two colors remain)

After the second checkmate, the game leads to the third and final checkmate. The following figure illustrates the arrangement of the pieces at the end of the game:

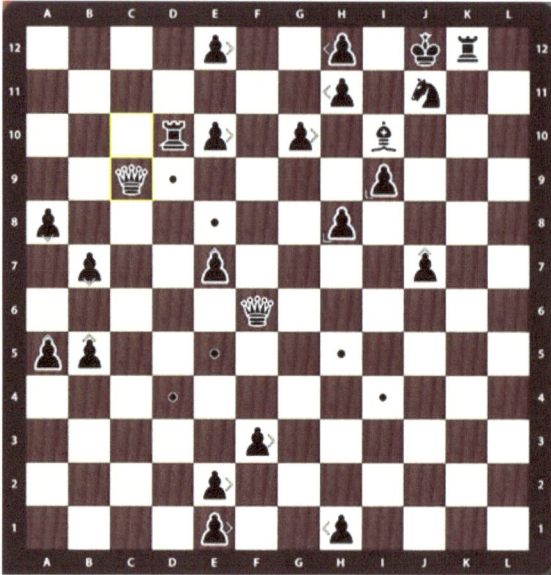

The final arrangement after the third checkmate

The final state on the board is like a union, whereby all remaining pieces become of the same color. Now they are all black, even though some were originally white, red, and green. They have been transformed into a new species. Or, we could say that all pieces have been transmuted into a higher form. In that new form, they have gained freedom because now they have access to the entire board, i.e., to those squares that were outside of their reach at the beginning of the game. Through their transformation, they have gained an extended awareness of their environment.

So, how does this relate to "a bishop changing its color"? How could the changing colors mark "the next revolution in physics"?

By changing their colors, the pieces demonstrate their "complex" structure, the structure of matter that is described in the previous chapters of this book. The quaternity pieces are "complex" because they appear in two colors, i.e., "real" and "imaginary." Their initial colors are the "real" colors; they correspond to the colors of their initial ordinary states. However, when they change colors, they manifest their "imaginary" parts.

The changing of colors by the pieces represents the "complex" structure of physical matter. As shown earlier, physical matter is also "complex"; it consists of "real" and "imaginary" parts. It is the discovery of this feature of matter that will lead to a revolution in physics. It is quite extraordinary that Feynman was able to predict that the discovery of the "complex" nature of matter would revolutionize physics!

***

The quaternity board determines the environment of the pieces. From the perspective of the pieces, there is nothing outside of the board; their "life" is confined to the board; the board is their universe. For an observer of the game who sees the board but is not aware of the presence of the players, it may seem that the pieces are aware of where they can go and where they cannot, that each piece has a unique ability to move. The important thing to notice is that an … invisible field drives all action on the board. This field controls and influences the entire dynamics of the game. There is no direct communication between the pieces. Instead, they all are connected or "entangled" together through this invisible field. They are linked together in the same way as the sand grains are connected through the acoustic field in the Chladni experiment. In the same manner, electrons behave within the structure of an atom, DNA molecules act within biological systems, and those huge galaxies rotate within their clusters. And this is the crucial observation to be taken from the comparison between quaternity and quantum mechanics. It may help us understand better the nature of quantum entanglement. Namely, entanglement is a form of manifestation of a template within which the pieces or particles are linked together. In other words, entanglement is the manifestation of a "higher-order" template which controls the pieces. It is that template which provides the pieces with their seeming awareness of themselves and of their environment. It is the template design which rules over all the actions. This is nicely illustrated in the following tale:

> Unjustly imprisoned, a tinsmith was allowed to receive a rug woven by his wife.
> A few days later, the man said to the prison's guard:
> "I am poor, and you are poorly paid. But I am a tinsmith. Bring me tin and tools. I shall make small artifacts which you can sell in the market and we will both benefit."
> The guard agreed to this, and they were both making a profit.
> Then, one day when the guard went to the cell, the door was open, and the tinsmith was gone.
> Many years later, when the tinsmith's innocence had been established, the man who had imprisoned him asked him how he had managed to escape; what magic he had used.
> The tinsmith explained:
> "It is a matter of design and design within a design. My wife is a weaver. She found the man who had made the locks of the cell door and got the template of the design

from him. This she wove into the carpet. After studying the carpet, I recognized the design. Then, I came up with the plan of the artifacts to obtain the materials to make the key – and that is how I escaped."[49]

As we can see from the story, familiarity with the design allows one to "open the door," change a bishop's color, or resolve the problems of modern physics.

***

As indicated earlier, the Chladni patterns represent only partially the acoustic field. They are two-dimensional cross-sections of a three-dimensional field. Therefore, what one can see in the Chladni patterns is only a tiny fragment of a much more sophisticated structure.

Similarly, one can look at the quaternity board as a two-dimensional cross-section across a multi-dimensional template that contains a fuller representation of the game. The quaternity template cannot be calculated by using elaborate numerical methods. Instead, the quaternity projection is determined by the perceptivity of the players. It is important to realize that all four players aim at the same overall objective: to transform all the pieces into a single-color set. This is the goal of all of them. Although each player may have a different approach and apply different strategies, the outcome is always the same: to end up with a united set of entangled pieces. From this perspective, it does not matter which of the players is the winner. In every game, the goal is accomplished.

---

[49] Adapted from the story entitled "The Design" included in *Thinkers of the East* by Idries Shah, The Octagon Press, London, 1986, p. 176.

At this point, it is interesting to quote Viktor Korchnoi, a chess Grand Master. He summarized quite precisely the binary limitation of traditional chess:

> The human element, the human flaws, and the human nobility – those are the reasons that chess matches are won or lost.

When the player's attention is driven only by the desire to win, it limits the experience to those two outcomes mentioned by Korchnoi. Such a binary mode of mind operation is limited to either "win" or "lose." Here we may see the same tendency that has been described in cases where people are trying to guess a card or dice. They will make so many mistakes that it is statistically impossible for them to be wrong so many times. This may be explained by the mechanism whereby the desire to "win" interferes with the more subtle ability to perceive the entirety of the game. In this context, "the human nobility" that Korchnoi refers to applies only to the ordinary, i.e., cruder mode of mind operation. Another form of nobility is within reach of the human mind. But to experience it, other layers of perception would have to be involved.

Quaternity allows exercising this other mode of the mind's operation, which may be referred to as perceptiveness. However, contrary to a common belief, perceptiveness does not mean delegating the leading role to the brain's right hemisphere. Rather, it means bridging the brain's two modes into harmonious cooperation. The harmonious cooperation of both hemispheres leads to the activation of higher modes of the mind. This means that it is not effective enough to look at the board only from one's own perspective, from one's own point of view. One must bring his or her mind to operate onto another level, i.e., onto a level from which it is possible to look from "above" at the overall picture that includes all other players' perspectives. This means being able to "see" the overall matrix that encompasses all four perspectives or the six elements of the game at the same time. The beginners try to imitate such an approach by continuously switching and scanning through all six elements of the game and then applying the acquired information to their situation. But such continuous switching and scanning is not effective enough. There is a much more effective way. However, it takes extra effort to develop such a more effective mode operation of one's mind. Quaternity offers the means to exercise such an ability. Of course, this would require acquiring special skills or developing not-so-ordinary perceptivity. This is schematically illustrated in the following figure, where the blue circle represents the overriding template of quaternity.

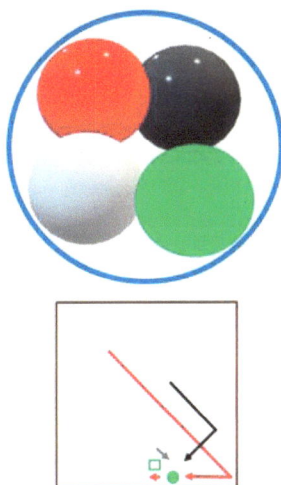

Overall template of quaternity

Now it is easier to understand that quaternity introduces some elements of what is referred to in this book as perceptive science, i.e., recognizing intimate interconnections which are perceptible only with reference to the whole. In this case, the "whole" means the "blue" circle within which the minds of four players are encompassed.

In this context, one may look at quaternity as a game which allows one to exercise the skills needed for dealing with such situations like the one facing today's physicists when trying to solve the mystery of "nothingness" and dark matter. Namely, an entirely new approach is needed. Symbolically, such an approach may be referred to as "seeing" as different from "knowing." This is illustrated in an episode describing a meeting between a philosopher and a mystic:[50]

> Ibn Sina, the great philosopher, met Abu Said. When they were asked to comment on the meeting, the philosopher said of Abu Said:
> "What I know, he sees."
> Abu Said said of the philosopher:
> "What I see, he knows."

While the philosopher refers to "knowing" as a deterministic approach, the mystic refers to "seeing" as a perceptive approach. The main point of the above exchange is that these two modes of comprehension, i.e., knowing and seeing, are not identical. The mystic sees what the philosopher knows. But this does not mean that the philosopher knows everything that the mystic can see.

---

[50] *A Perfumed Scorpion*, Idries Shah, The Octagon Press, London, 1978, p. 69.

The relationship between regular chess and quaternity is similar, chess is like "knowing" but quaternity is "seeing." As demonstrated by the computer's ability to outperform the best human players, regular chess is confined to the deterministic or binary mode of thinking. On the other hand, quaternity provides an opportunity to "see" the available options and respond more effectively. This kind of "seeing" the game is partially equivalent to that described by Mozart when perceiving music.

A similar form of "seeing" has also been experienced by the greatest artists and scientists. An example of an artistic representation of such "seeing" is the painting "Burial of St. Lucy" by the Italian painter Amerighi da Caravaggio.

"Burial of St. Lucy" by Caravaggio (photo credit: Dominique Hugon)

The painting is on display in the Basilica Santa Lucia al Sepolcro in Syracuse, Sicily. Painted in 1608, it shows a scene from the burial of Saint Lucy. The scene is set in a dark crypt. At the front, there are two muscled gravediggers. The mourners are much smaller; they seem to be placed some distance away. The cavernous space of the crypt dwarfs all those present there. It is the overwhelming dark crypt which contains the crucial element of the painting. Namely, if we take a closer look at the upper part of the painting, then we can recognize that two-thirds of the space is filled with the image of a woman's face. The face is embedded within the walls of the crypt. It is the woman's face that encompasses the entire scene.

It may help to grasp the scene's meaning to compare it with the illustration of the template of quaternity. Just like the blue circle which contains the template of the quaternity field, the woman's face encompasses everything that is taking place in the crypt.

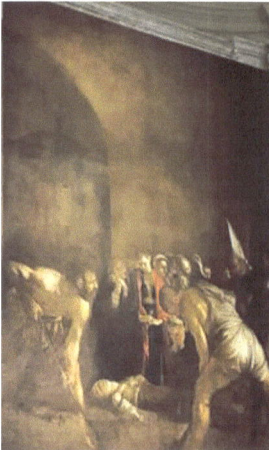

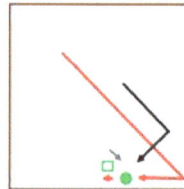

Two templates

Another example of such "seeing" was experienced by Henri Poincaré, a French mathematician, theoretical physicist, engineer, and philosopher of science. As a mathematician and physicist, he made fundamental contributions to the research on the three-body problem. This example is taken from one of his greatest discoveries, the first which consecrated his glory, i.e., the fuchsian theorem. (It may be helpful to quote Poincaré's own precaution: "This theorem will have a barbarous name, unfamiliar to many, but that is unimportant.")

> Just at this time, I left Caen, where I was living, to go on a geologic excursion under the auspices of the School of Mines. The incidents of the travel made me forget my mathematical work. Having reached Coutances, we entered an omnibus to go some place or other. At the moment when I put my foot on the step, the idea came to me, without anything in my former thoughts seeming to have paved the way for it, that the transformations I had used to define the Fuchsian functions were identical with those of non-Euclidian geometry. I did not verify the idea; I should not have had time, as, upon taking my seat in the omnibus, I went on with a conversation already commenced, but I felt a perfect certainty. On my return to Caen, for conscience's sake, I verified the result at my leisure.[51]

What is extraordinary about this episode, and at the same time relevant to our discussion, is that this incredibly sophisticated idea came to Poincaré in a flash. Roger Penrose commented on this episode in the following way:

---

[51] *An Essay on the Psychology of Invention in the Mathematical Field*, Jacque Hadamard, Princeton University Press, 1945.

It should be made clear that the idea itself would not be something at all easy to explain in words. I imagine that it would have taken him something like an hour-long seminar, given to experts, to get the idea properly across. Clearly, it could enter Poincaré's consciousness, fully formed, only because of the many long previous hours of deliberate conscious activity, familiarizing him with many different aspects of the problem at hand. Yet, in a sense, the idea that Poincaré had while boarding the bus was a 'single' idea, able to be fully comprehended in one moment! Even more remarkable was Poincaré's conviction of the truth of the idea, so that his subsequent detailed verification of it seemed almost superfluous.[52]

So, let us go back to quaternity. Instead of switching on and off from a specific to a general situation of the game, a player could "see" the game from a much more comprehensive perspective. This does not mean that such a player would always be able to win. Of course, each player should do his or her best and, in this way, contribute to the richness of the game. However, what matters here is the … experience – just like in the case of "conceiving" an image, or "hearing" compressed music, or "seeing" a mathematical solution. Whether such "conceiving," "hearing," or "seeing" leads to winning, fame, or richness – is irrelevant. What is important is the very experience of such extraordinary moments. In quaternity, such moments appear quite frequently. It could be your move or a move by another player that, at first, may seem irrelevant, but a few rounds later, turns out to be critical to the overall outcome of the game. From this perspective – it does not really matter who

[52] *The Emperor's New Mind*, Roger Penrose, Oxford University Press, New York, 1989, p. 419.

wins the game. Instead, there is the opportunity to practice such a much more advanced sort of "seeing." Therefore, what matters is to be able to "hear" the game, to "feel" how it unfolds, and "see" what opportunities become available to each of the players. And that's the greatest reward – to be able to perceive the game in its entirety. It is a form of experiencing a certain kind of "beauty." It is the same sort of experience as that of the Troubadours when they "sighed for the love of a lady," or the "beauty" of the woman in Caravaggio's painting, or that which is referred to as "heavenly music" or "heavenly harmony." For example, Shakespeare delegated Lorenzo in *The Merchant of Venice* to describe such an ability:

> There's not the smallest orb which thou behold'st
> But in his motion like an angel sings,
> Still quiring to the young-eyed cherubins;
> Such harmony is in immortal souls;
> But whilst this muddy vesture of decay
> Doth grossly close it in, we cannot hear it.
> (*The Merchant of Venice*, V.1)

Now it is easier to see how quaternity is a tool that may help to deal with "this muddy vesture of decay" which prevents one from "seeing" or "hearing" or "touching" or "tasting" or "feeling" this kind of beauty.

What is important for our discussion is the realization of the existence of a higher-level structure within which these various forms of beauty are embedded. As seen in the game of quaternity, such a structure is multi-layered. What is available to ordinary senses and rationale – is limited to the lower level. In order to understand the behavior of things, it is necessary to perceive their overriding template.

And this brings us to the following conclusion: to resolve the current impasse in science, the overall framework will have to be expanded. Such an expansion will require the templates from which the "complex" patterns are projected onto spacetime. This means that the familiar deterministic worldview will be augmented by an entirely new set of forces and "complex" particles.

Science of Consciousness

> The universe is a gradient of consciousness and on this gradient the earth occupies a low level. Its highest raw material is mankind.
>
> *(Ernest Scott)*

We can now put together the structure of the universe. We have considered three major regions of the universe, each represented by a pentangle, i.e., matter, the Earth, and the human mind.

The three pentangles together illustrate the complete structure. On the following diagram, the lowest pentangle represents matter, the second (middle) pentangle represents the Earth, and the top one represents the human mind.

Let's add some details which may help to grasp the overall concept illustrated here. The lowest pentangle on the diagram represents the various layers of physical matter, starting with elementary particles. The blue contour around the external star indicates the boundary between "nothingness" and quantum particles.

The structure of the Earth is shown as the pentangle in the middle. The various layers in this pentangle correspond to the lithosphere, hydrosphere and atmosphere, biosphere, fauna, and mankind.

210

Structure of the universe: matter (bottom); the Earth (middle); the
human mind (top); the blue contour represents the boundary
between the quantum world and "nothingness;" the green contour
marks the boundary between physical senses and subtle faculties.

The pentangle on the top is a view of the human mind. The
largest star in this pentangle corresponds to the physical senses.
The physical senses constitute the upper limit of the physical
world. The green contour marks a buffer zone, i.e., the
boundary between physical senses and subtle faculties. The

buffer zone veils an ordinary human mind from access to subtle perception, i.e., perception of the invisible worlds. The physical world is enclosed between the blue and green contours.

The important thing is that every layer within the physical world has a template placed in the invisible world. This is shown in the following illustration.

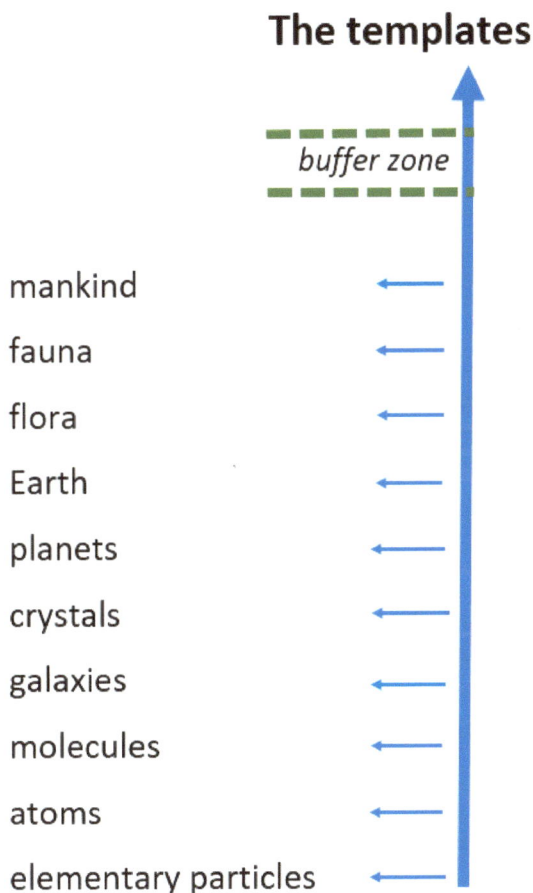

The templates

buffer zone

mankind ←

fauna ←

flora ←

Earth ←

planets ←

crystals ←

galaxies ←

molecules ←

atoms ←

elementary particles ←

The gradient of consciousness
(the horizontal arrows indicate the links with the corresponding templates of the invisible world)

In this illustration, the vertical arrow indicates a gradient of complexity along which the various forms of matter are fashioned. The horizontal arrows show the links between physical objects and biological systems and their corresponding templates. Now we may realize that the templates perform the role which quantum mechanics assigned to an "experimenter." It is the templates placed outside of the physical dimensions that enforce the shapes of physical objects and biological systems.

This illustration helps to reconstruct the various stages of the evolution of these different forms of matter and their gradual appearance within spacetime. It is impossible to construct a theory of matter without these templates. Neither is it possible to describe the evolution process satisfactorily if these templates are not included. This is because the templates are the prime drivers of the entire evolution.

The model of the universe described here provides a framework within which it is possible to put together hard-core physics with consciousness. Moreover, it is this framework which is needed to define consciousness. It is this framework that was missing from all the various descriptions that are quoted at the beginning of this book.

The structure's main feature is that every element has a certain degree of awareness. Every particle, object, or biological system has a sensing apparatus, which defines the extent of its awareness. Secondly, as we go up along the gradient, these various objects and systems acquire increasingly more sophisticated sets of sensors, senses, and faculties. It is the quality and sophistication of the sensing apparatus which determine awareness. The expansion of awareness translates into an expanded environment which a given object or system can perceive and respond to. Thirdly, the boundary of the environment within the grasp of each object or system is determined by the corresponding template in the invisible world.

Now we are in a position to define what consciousness is. Namely, consciousness is awareness of the environment by a given object or system through a set of senses and faculties. Therefore, consciousness is determined by the corresponding template of the given object or system. As we go along the spectrum of matter, the degree of consciousness becomes more and more sophisticated. This means that consciousness is relative; it depends on an object's position within the overall spectrum of matter.

For example, electrons are conscious of the structure of the atom they are part of. Molecules are conscious of the overall symmetry of a crystal. The molecules of DNA are conscious of the physical form of the systems they build. The same applies to planets in planetary systems and galaxies in galactic clusters.

When we reach the level of flora and fauna, a much more advanced sensing apparatus becomes available. For example, plants can sense minerals, water, heat, temperature, light, etc. Animals are equipped with additional faculties that allow them to expand their awareness of their environment. In addition to the five physical senses, humans are equipped with several faculties, such as memory, imagination, fantasy, talents, and ego. These further enrich their awareness of their environment, which include physical surroundings as well as a range of feelings, sensations, mental images, and psychological states.

Each system, however, is aware only of its immediate environment. The environments that are one or more steps higher within the overall structure of matter – are beyond its reach. For example, electrons are not aware of the symmetry of crystals, nor atoms can perceive the structure of biological systems. Neither ordinary humans are aware of the invisible realms.

The crucial point is that the templates are the ultimate source of consciousness. The templates determine what senses and faculties are available to an object or a system. In this context, the templates are the actual minds:

> Water and salt and minerals and elements,
> all mixed up and held together
> by this thing you call your mind.[53]

The imaginary mates of the quark-like particles facilitate the projection of the templates. These imaginary parts permeate every physical object and biological system. Every object, system, or human body contains innumerable amounts of these imaginary parts. The imaginary parts form something that could be called the "soul" of those systems. And it is these "souls" that provide access to consciousness. In the case of humans, consciousness is transmitted through every single cell of the human body. This is the biggest secret of nature: Nature is "complex," every speck of matter contains a soul. This is nicely expressed by Gharib Nawaz[54] in the following quote:

> Every particle of dust is a cup wherein
> all the world can be seen.

It is now possible to understand physicists' difficulty when constructing the "theory of everything." This theory is supposed

---

[53] *The Mines of Light*, Arif Shah (see note #42).
[54] Gharib Nawaz (Benefactor of the poor) is the title given to Moinuddin Chishti, an Indian mystic (1141-1230).

to describe the relationships between all physical objects without complex particles, i.e., a model in which there is no room for "souls." Such a theory is ... an impossible task.

Another essential feature of the overall structure should be noted and emphasized. Namely, all the templates are projected onto the physical world in complete forms. This applies from the tiny elementary particles up to the galaxies, flora, and fauna. However, when the templates are projected onto spacetime, they are slightly skewed during that process. There is a small margin for some "blurring" of the original templates. Therefore, this blurring leads to diversity within the same species groups. This is illustrated in the following story:

> A man was ill and, although apples were out of season, he craved one.
> Hallaj, the wise one, suddenly produced one.
> Someone said: "This apple has a maggot in it. How could a fruit of celestial origin be so infected?"
> Hallaj explained:
> "It is just because it is of celestial origin that this fruit has become affected. It was originally not so, but when it entered this abode of imperfection, it naturally partook of the disease which is characteristic here."[55]

The universe is like a huge operating machine. All objects and systems are entangled through the templates of the invisible world. All the templates form what could be called the Entangled Mind of the universe. Or, as perceived intuitively by Max Planck, "this mind is the matrix of all matter."

---

[55] "The celestial apple" from *The Way of the Sufi* by I. Shah, Octagon Press (London, 1968).

The Entangled Mind controls this machine. Within it, all templates are linked together. They all together form a single entity. Therefore, if needed, some changes can be made. For example, the re-arrangement of gigantic clusters of galaxies located many light years away from the Earth can indirectly induce volcanic eruptions which, in turn, can make needed changes within the biosphere.

In summary, it is consciousness that drives the entire mechanism of the universe. The inclusion of consciousness into the model of matter allows to solve the current challenges of modern physics outlined in this book. A sort of conceptual quantum leap will have to be made to accept the "complex" model of the universe, which is needed to advance the current state of science.

It will be interesting to see when and in what manner such a complex model of the universe will be "discovered" by physicists. When we start to see headlines in newspapers and social media announcing that scientists have discovered a family of complex particles, this will mark the opening of a new chapter in modern physics. This will be the beginning of the revolution to which Richard Feynman referred. This may happen tomorrow, next year, or in the next decade. In the meantime, we would better get used to seeing such headlines as "the universe is a hologram," "we live in parallel universes," "the multiverse is the answer," etc., etc. These headlines indicate that physicists are still locked up in their religiously deterministic world.

There is one more aspect of the model of the universe that should be mentioned. It does not apply directly to hard-core physics, but it may help to grasp the overall concept. This aspect is relevant to the science of the human mind. Namely, there is one exception to the completeness of the universe that was mentioned above. The universe operates smoothly, but one system appeared in an incomplete form. One system is non-mechanical. This exception is ... mankind. Yes, that's correct. The template that was used to project mankind onto the physical world was not quite complete. The result of this partially incomplete projection is that the environment that people can perceive is not quite adequate to their situation. Despite all their senses and faculties, people are incapable of being fully aware of the environment which is within their reach, and which affects them. They are missing faculties that would allow them to be aware of their potentiality and the function they are supposed to discharge. Previously, we referred to these missing faculties as the subtle faculties that remain latent in ordinary men and women. Now we may refer to these faculties as a subtle "soul" needed to cross the buffer zone between the visible and the invisible. Without it, people cannot grasp the full extent of their role within the overall structure. As a result, people feel that they are incomplete:

> The unattainable lady of the Troubadours symbolized a quality to which man might feel drawn, but which was essentially inaccessible to him in his ordinary state.[56]

If we use the terminology quoted in the previous chapters of this book, the subtle "soul" may be compared to the missing precious "pearl" or to "the child of the spirit":

> The man who knows must be aware that
> the child of the spirit is born in one's heart.

---

[56] *The People of the Secret*, Ernest Scott (see note #8).

Although the subtle faculties are latent, they can occasionally "leak" some faint signals of their presence. And it is these occasional fleeting moments that make people feel unsatisfied, confused, and unhappy. Why is it like this?

Because the human potential belongs to the higher worlds. Humanity belongs to the timeless and placeless realm. Unlike the other systems, humans have been granted the potential to move up along the gradient of consciousness, all the way to the invisible worlds and beyond. The human mind is capable of reaching and penetrating the invisible worlds. However, this ability is not given; it is not automatic. It remains in its latent form. A very sophisticated methodology is required to activate this potentiality. It is in this sense that humankind may be referred to as the "highest raw material" of the universe.

***

The entire universe has been created to allow men to explore themselves. Man's role is to sustain the universe and develop further the invisible worlds. Man's function is to contribute to the development of the New Cosmos. Therefore, there must always be a few individuals capable of existing simultaneously in this physical world and in that timeless and placeless realm. They provide and preserve the link between the invisible and the visible worlds; they function as the "attachment points" to the highest zones within the invisible realms. In this way, they sustain the entire universe. Regardless of how unlikely or even impossible this may sound – it does not change the fact that the physical universe would cease to exist if there was no one to provide a link to the invisible realm. Such a link is the critical part of the Entangled Mind.

Books by the same author

The New Cosmos, Troubadour Publications (2021)

A Journey through Cosmic Consciousness, Troubadour Publications (2019)

A Journey with Omar Khayaam, Troubadour Publications (2018)

Shakespeare's Elephant in Darkest England, Troubadour Publications (2016)

Shakespeare's Sequel to Rumi's Teaching, Troubadour Publications (2015)

Shakespeare's Sonnets or How heavy do I journey on the way, Troubadour Publications (2014)

Shakespeare for the Seeker, Volume 4, Troubadour Publications (2013)

Shakespeare for the Seeker, Volume 3, Troubadour Publications (2013)

Shakespeare for the Seeker, Volume 2, Troubadour Publications (2013)

Shakespeare for the Seeker, Volume 1, Troubadour Publications (2012)

En español

El nuevo cosmos, Editorial Sufi (2021)

Un viaje por la consciencia cósmica, Troubadour Publications (2020)

Un viaje con Omar Khayaam, Editorial Sufi (2020)

Shakespeare para el buscador (Completo: 4 volúmenes – versión Kindle), Editorial Sufi (2020)

El elefante de Shakespeare: en la Inglaterra más oscura, Troubadour Publications (2017)

Rumi y Shakespeare, Editorial Sufi (2016)

Shakespeare y su maestro, Editorial Sufi (2015)

Shakespeare para el buscador - Volumen 4, Editorial Sufi (2013)

Shakespeare para el buscador - Volumen 3, Editorial Sufi (2011)

Shakespeare para el buscador - Volumen 2, Editorial Sufi (2011)

Shakespeare para el buscador - Volumen 1, Editorial Sufi (2011)

En français

Voyage à travers la conscience cosmique, Troubadour Publications (2021)

www.ingramcontent.com/pod-product-compliance
Lightning Source LLC
Chambersburg PA
CBHW050438240326
41599CB00060B/8